PIET OUDOLF AT WORK

10 × 10

Toolsmith
16'/5M
POSCA
Sharpie

Introduction: Piet Oudolf and the Zeitgeist in Planting Design

Cassian Schmidt

There is an amazing sense of togetherness among garden designers – an atmosphere of respect, appreciation and mutual inspiration. And yet each one of them has developed their own authentic signature. In 2019, I had the privilege to join Piet Oudolf and other leading designers, such as Claudia West, Tom Stuart-Smith, Thomas Rainer, James Hitchmough and John Greenlee, on an unforgettable field trip to Sichuan, China. We all love nothing more than to talk about plants and share our enthusiasm for nature. Studying plant combinations in nature and experiencing familiar or new perennials in their natural context inspires us. Piet is uncomplicated on such trips, and despite being almost 75 at the time, happily put up with the long car journeys, the mountain passes over 15,000 feet and the often simple accommodation in China.

Piet is a morning person, an early riser. During our time in Sichuan he and I arranged to meet now and then before breakfast to experience the special early morning atmosphere in the mountain villages, with their beautifully painted Tibetan houses. Piet has a knack for quietly enjoying and absorbing moments like this without the need for many words. He also usually goes out early in the morning at his garden in Hummelo, to experience the special atmosphere then and capture the development of his plantings in photographs. His atmospheric images of autumnal plantings shrouded in dew and morning mist, published in his first books, were influential in terms of style at the time, and undoubtedly contributed much to the idea of a new way of looking at plants, one that was less about flower and more about form. Piet has a keen sense of what gives a plant its character, and which plants may offer potential for the future. Over the years we have selected some useful plants during his annual visits to Hermannshof. Back in the 1980s he had already selected and named numerous plants for his own projects at his nursery in Hummelo.

Despite his celebrity, Piet has kept his feet on the ground. He remains an approachable individual who feels especially at home among like-minded people. I am always amazed to see him patiently answering questions from students, whenever he encounters young people with a special enthusiasm for plants at conferences. He shares his knowledge, experience and planting philosophy openly and generously with others. His planting plans and plant lists are not secret, unlike with most other designers; he publishes them both in books and online. Piet is surprisingly relaxed about the possibility that others might copy him, saying they are welcome to do so. But by the time any given design has been completed, Piet has long since moved on in his creative thoughts and visions, and never reuses exactly the same combination of plants. In recent years, when he might have eight or ten projects to oversee and install simultaneously in different corners of the world, Piet has sought out a handful of collaborators to implement his projects – partners he trusts absolutely to coordinate the implementation of his planning in the way he intended.

***

Contemporary planting design in urban open spaces involves balancing the needs of ecology, aesthetics and maintenance. Current designs focus on the way in which the space is felt and experienced as a piece of nature, and unusual plant compositions aim to attract the attention of 'users'. The focus is on achieving an overall impression that, while clearly having been created by planners and gardeners, at the same time is reminiscent of a certain natural model. Against the ever-present backdrop of climate change, issues such as urban climate and ecology, biodiversity and the economical use of resources are all increasing in importance. Reducing installation and maintenance costs is another determining factor when it comes to providing a rich and rewarding experience of perennial planting design.

Never before has there been such an incredible array of interesting plant species and varieties available to landscape architects and designers as in the last 20 years. Precious little of this can be seen in most contemporary projects, however. The necessarily nuanced approach to planting design would appear to be so

For me garden design is not just about plants, it is about emotion, atmosphere, a sense of contemplation.

Piet Oudolf

complex that only a few well-versed specialists dare to attempt it. These few outstanding designers have had a significant influence on planting design, especially since the 1990s, and each of them has advanced the discipline in their own way. These individuals make up an astonishingly small group, whose members all know each other well and maintain a productive relationship through encounters at international conferences or in collaborations. In addition to a profound horticultural background, they all have in common a great creative flair. This actually makes them the exotics of the design world, because today's landscape architects often have very little basic horticultural knowledge – which explains why we see plants being used in superficial and unimaginative ways in many projects.

The term 'New Perennial movement' has been used to describe a trend in contemporary planting design that has emerged in Europe and North America over the last 20 years; it focuses on perennials as its strongest stylistic element. For me, the 'movement' element is a slight exaggeration, however, perhaps even misleading – there is no manifesto or formal association emanating from those involved. What they *do* represent is, rather, a common direction or mindset in planting design, combined with an expression through plants that is more naturalistic in some cases, less so in others.

Alongside the development of a contemporary, habitat-based, but at the same time more artistic approach to planting design by Piet Oudolf came parallel developments in both the UK and Germany from the 1990s onwards. While the common denominator for all those involved in the movement is the predominant use of perennials appropriate to the location, the planning approaches vary significantly in the extent to which dynamic development is permitted – or not, as the case may be. The key question that designers of perennial planting schemes must ask themselves from the outset is, will my design be allowed to evolve continuously through self-seeding and natural renewal, without too much intervention by the gardeners? Or am I more concerned about preserving certain proportions and distribution patterns, with a view to maintaining and further developing the overall artistic impression? Associated with the dynamic planting principle is also a generally more academic approach, in which vegetation science, plant sociology and ecological insights play an essential role.

The UK's Sheffield School has represented this approach since the late 1990s, through the key figures of Nigel Dunnett and James Hitchmough from the Department of Landscape Architecture at the University of Sheffield. Both designers incorporate elements such as the Plant Survival Strategies of plant ecologist John Phillip Grime (1935–2021) into their planting and sowing concepts. Nigel and James came to my attention through their highly mixed meadow-like plantings composed of various layers at London's Queen Elizabeth Park, created for the 2012 Summer Olympics. In the UK these naturalistic concepts are referred to with terms such as 'enhanced' or 'stylized nature'. Both in the UK and in Germany, planting is not limited to native species, but comprises combinations of native and exotic species. This naturalistic approach to achieving an enriched and enhanced impression of nature in the garden is not new, but harks back to William Robinson's *The Wild Garden* (1870). In Germany at the beginning of the twentieth century, Willy Lange (1864–1941), among others, presented a naturalistic style orientated towards the physiognomy of the plants – their 'character', as Piet would say.

The German Hansen School was established much earlier than the Sheffield School, as an approach to planting design that focuses on plant sociology while simultaneously emphasizing the aesthetic. This concept is based on the garden habitats and levels of plant sociability developed by Professor Richard Hansen and Hermann Müssel in Weihenstephan in the 1970s; these have strongly influenced planting design in Germany since the early 1980s. The British plantsman and garden writer Stephen Lacy coined the term 'New German Style' for this approach to applied plant sociology in 2002, having seen important examples of the innovative style on his travels through Germany a decade earlier. Lacy was impressed by the naturalistic, steppe-like plantings of Rosemarie Weisse in the Westpark in Munich, the generous perennial areas designed by Urs Walser at the Killesberg garden in Stuttgart, and the wild perennial plantings at Hermannshof in Weinheim, also designed by Walser.

Piet had visited these pioneering gardens in the 1990s. He was especially inspired by the abundant use of drought-tolerant grasses from steppe and prairie habitats, having not seen anything comparable in the Netherlands at the time. Yet he never went on consistently to apply the habitat concept in his planting design, finding the stringent list-based approach ultimately too restrictive and academic. He has always preferred to rely on his many years of empirical gardening experience - which has ultimately led to very similar results. This is undoubtedly the key to the success of Piet's plantings. He stays true to his horticultural roots, masters his plant material like no other, knows exactly what can be combined and what cannot, what will stay the course, and how the plants will behave in the longer term. At the same time, however, he has the gift of combining plants to great effect, with an outstanding sense of rhythm, proportion and, above all, the temporal dimension - the one that fundamentally distinguishes garden art from fine art.

***

This book looks at Piet Oudolf as a creative practitioner, and at the development of his work's appeal for a wide range of audiences, from city-dwellers to private gardeners to landscape designers and horticulturalists. It is also the first book to present an overview of his planting plans. Piet is surely the first garden designer to have plans shown in his own exhibition at the Hauser & Wirth gallery in Somerset in 2014; in spacious rooms otherwise reserved for contemporary artists, the drawings were elevated to the status of art object. This is astounding when we recall that the planting plans were never intended as art, but as a tool through which to visualize his thoughts and put his ideas into practice. The drawings, the coloured plans in particular, have a unique graphic quality and aesthetic appeal that allow them easily to stand on their own, apart from the gardens they symbolize. This is due, not least, to the fact that they bring to mind an abstract image, being made up of a multitude of elements and coloured signs that are hard to interpret at first sight. For the professional, they also serve as precise technical instructions that accurately represent everything necessary to implement the plan. The documents are executed by Piet himself, and no one else. He may well be the only designer who does all of this stage of the work himself, including the task of calculating plant quantities. In recent years, however, he has yielded to the need to delegate the procurement of the plants to a Dutch office. Also, with the exception of a few trainees, Piet has never employed any permanent staff. His wish has always been to maintain his independence and, above all, avoid the burden of being financially responsible for anyone else.

How have his plans changed in his 30 years as a designer? Initially - until around the end of the 1990s - they were almost exclusively drawn in black and white. This more technical representation was in line with the reduced, one-dimensional plantings and the often rather formal design language of the time (pp.48-9). Later, from around 2000 onwards, Piet's planting schemes became increasingly complex, varied and, above all, naturalistic. He began to incorporate different layers and functional types in his plantings, such as the 'scattered plants' and 'individual plants' marked with specific symbols in his plans. This approach called for more complex, coloured planting plans to make them easier to read and to implement on site; indeed, at the Vitra project in Weil am Rhein (pp.202-19), the planting was done using a grid and coloured spray paint on the ground (see also 'Planting the Vitra Garden', p.210).

Once he had multiple projects on the go at the same time, Piet was no longer able to handle the physical distribution of the plants in person. This made it all the more important to specify each element graphically in his plans, and to provide information on the exact distribution of the plants in the key, which he structures according to the various categories of use. He develops his plans with different layers on tracing paper, each of which can be viewed separately but can also be laid on top of each other to create an overall image.

Piet's plantings from recent years are relatively species-rich, with an average of 80 to 120 in total, but sometimes over 200. I only realized how complex his plans really are when I spent a day helping to arrange the perennials at the Vitra garden. The sheer quantity alone - over 36,000 of them - called for excellent coordination by everyone involved and, above all, precise allocation to specific areas - combined, if possible, with a good knowledge of the plants to avoid confusion.

Students and landscape architects can learn a great deal about well-functioning plant combinations and seasonal progression from Piet's plans; indeed, this is one of the key aims of this book. The collection of plans presented here serves to demonstrate his clearly identifiable and unmistakable artistic signature, while also helping to illustrate the masterful refinement of the details of his plans over time. This unmistakable signature sometimes gives rise to the accusation that Piet's work has developed little in recent years, and that the plants are always typical Oudolf choices. This may well be true at first glance, for it is the recurring elements such as prairie grasses, umbellifers, *Sanguisorba* (great burnet) and certain muted hues that make his work so recognizable. But on closer inspection, the attentive observer will see the constantly evolving plant combinations and the many innovations in detail, and conclude that the designer's repertoire is by no means limited to specific combinations. Looking at the whole range of plans from the last 30 years, we see a clear development towards more naturalness and complexity (see, for example, pp.144, 176, 182).

Initially, from the 1980s to the late 90s, Piet worked almost exclusively with classical, conventional mosaic and block plantings - for example at the Dream Park in Enköping, Sweden (1996; see pp.247, 249), one of his first larger-scale projects. In Piet's early designs, such block plantings, traditional in the UK, consist of only one species at a time, usually covering an area of around 3 square metres (*c.*32 sq. ft). The essential difference in respect to traditional British plantings, however, was that even then Piet used plants with a wild and natural character, plants that retain their aesthetic quality as dry skeletons in winter.

From around 2000, Piet started developing mixed blocks and mixed plantings, such as at the Lurie Garden in Chicago (completed 2004; see pp.78-9) and on New York's High Line (opened between 2009 and 2019; see p.91). A variation on this theme was his occasional use of mixed drifts or bands, such as at RHS Garden Wisley in the 1990s (see p.249, fig.7). His primary goal in combining species that complement each other in growth characteristics is to increase diversity in terms of the visual effect. These drifts or bands comprise at least two species, though in the case of Wisley they include five or six. Typical plant combinations that Piet has developed and perfected over years of experience include late-emerging *Cimicifuga* species underplanted with early emerging *Briza media*, *Selinum wallichianum* with *Geum* hybrids, *Cenolophium denudatum* with *Eragrostis spectabilis*, *Iris sibirica* with *Astrantia* or *Echinacea* with *Calamintha nepeta* 'Triumphator'. By combining about 60 per cent low-growing perennials with 40 per cent taller species that develop little foliage, he achieves an effective layering approach that helps to reduce maintenance.

Over the last ten years, Piet's work has been characterized by a much more naturalistic approach to matrix planting. I remember, during one of his visits to Weinheim, our detailed conversations about this new way of using plants, inspired by the natural vegetation of meadows and prairies. Back then, in the mid-2000s, there was no name for this new type of planting. Searching for a suitable term, we decided that 'matrix planting' would be appropriate. The term originally comes from geology and refers to rocks, such as porphyry, that consist of a fine-grained, uniform groundmass in which coarser minerals such as feldspar are embedded. Piet had used the principle once before, many years earlier, in a small area at Bury Court in Surrey (see p.49). He had created a matrix of *Deschampsia cespitosa* 'Goldtau'; later replaced by more durable *Molinia*, interspersed with *Digitalis ferruginea*, *Dianthus carthusianorum* and *Trifolium rubens*, and I had only seen the result in photographs. This use of perennials was completely new to me, and I was immediately inspired.

Matrix plantings consist of a dominant species that sets the tone (usually a grass like *Sporobolus*, *Molinia* or *Sesleria*), through which groups of plants are interspersed like islands, representing different flowering aspects throughout the year. The matrix planting at the centre of the Vitra garden (pp.212–15), which consists of 60 per cent *Sporobolus*, is embedded with rhythmically repeated and amorphous islands occupied by a range of different species. Typical plants used in larger groups in the islands include *Salvia pratensis* 'Rhapsody in Blue', *Monarda bradburiana*, *Eryngium* × *zabelii* 'Big Blue' and *Helenium* 'Moerheim Beauty'. The islands consist of species that flower mainly in early summer and are then deliberately not cut back, as the seedheads of these perennials go on to provide a completely different picture in summer, autumn and winter and form an important part of Piet's 'aesthetics of decay'.

The islands are supplemented by scattered plants, distributed as if at random, to create a dynamic impression. In the Vitra planting, for example, Piet used *Sanguisorba menziesii* 'Wake Up', *Echinacea pallida* 'Hula Dancer' and *Asclepias tuberosa*, all of which later stand out from the matrix with their attractive seedheads. He also distributes individual plants – either perennials like *Papaver orientale* 'Karine' that die back early, or short-lived species such as *Lobelia* × *gerardii* 'Hadspen Purple', which can never be used in larger groups. He knows exactly which plants serve which function in the planting scheme, and which distribution pattern they can be used in.

In plans from recent years, Piet often uses all three types of planting in one project at the same time. This hybrid style of blocks, mixed blocks and matrix plantings can be seen in its basic form for the first time in Chicago's Lurie Garden (pp.78–9). Piet went on to further develop the idea of matrix plantings and increasingly placed them at the centre of his designs, including in the garden at the Hauser & Wirth gallery in Somerset (pp.134–35), and in his 2020 creation for the Vitra Campus in Weil am Rhein (see pp.212–15).

Pruned woody plants and cloud-like trimmed mixed hedges played an important role both in Piet's own garden in Hummelo and in his early projects up to around the year 2000, and as living sculptures they served as an unmistakable contrast to the informal perennial plantings. But from around 2005 onwards, formal woody elements almost completely disappeared from his repertoire, with free-roaming, picturesque woody plants such as *Sassafras albidum*, *Cotinus obovatus*, and *Betula papyrifera* taking their place. At the Lurie Garden, and especially on the High Line (designed in collaboration with James Corner Field Operations), Piet took a very nuanced approach in his use of a broad spectrum of expressive, predominantly North American woody plants (see pp.82–3, 86–7, 90–92). The result is a series of natural impressions with different lighting moods, from savannah-like light stands of shrubs to small forests of birch trees, or lush tropical areas with large-leaved magnolias. In a private project at an estate on the island of Nantucket, Massachusetts, also designed in collaboration with James Corner Field Operations, woody plants play an essential role in the form of frame planting, providing a windbreak or shelter belt from ocean winds and spray.

Owing to the relatively one-sided choice of images used in most books, the generous perennial and grass areas are the best-known sections of Piet's published gardens, and this helps to explain why he is generally seen as a virtuoso practitioner with perennials, rather than a serious user of woody plants. This is a significant misunderstanding, he says, because he is, in fact, very interested in combining shrubs and perennials. Indeed, many of his projects over the last few years have involved at least as many shrub underplantings and shrub border plantings as pure perennial plantings in open, sunny areas.

One question I am often asked by students or colleagues in connection with Piet's planting plans and projects concerns naturalness, sustainability and longevity: are his plantings really natural, or are they just 'naturalistic'? Are they based on ecological principles, or are they first and foremost the result of an artistic vision? There are plenty of misinterpretations and misunderstandings in this regard, depending on the perspective and background of the beholder. If we analyse the plans and the planting schemes closely, we see that they are based, at least in part, on rather traditional patterns of arrangement - yet they manage to create a surprisingly naturalistic impression. This appearance must not be equated with the terms 'natural' or 'ecological', however. Piet's plantings, while strongly inspired by natural landscape models, are at the same time enhanced, scenically arranged and abstracted versions of nature. They are planned down to the last detail, and artistically composed.

Their momentum is controlled deliberately through targeted maintenance interventions, and Piet's plantings cannot be described as low-maintenance *per se*. Despite never having claimed this as one of the essential goals of his planting design, he is often measured against it. Well-trained staff are therefore key when it comes to maintaining his designs. Finding the appropriate staff is often a challenge, and Piet always takes maintenance into account during the planning phase, asking his clients who will be doing it and what qualifications they will have. Even after installation, he takes care of management through annual visits, and handles improvements in close coordination with the gardeners involved.

Piet's designs generally provide a good basis for at least reduced maintenance when compared to conventional perennial borders (20-30 minutes per square metre (*c.*10 sq. ft) per year). This is based on clearly readable planting schemes, use of plants that grow in clumps and rarely self-seed, use of long-lived, reliable and competitive species, a focus on plants with a broad range of light and soil requirements, use of high planting densities (up to nine per square metre for matrix plants) in order to achieve fast and, above all, dense vegetation coverage, as well as skilful vertical layering. This is achieved by combining low-growing, early-emerging or fast-developing 'nurse plants' with late-emerging taller plants.

The ecological aspects and sustainability of Piet's plantings are another issue. In contrast to the scientifically based concepts of the Sheffield School and the phytosociological approach of dynamic mixed plantings in Germany that represent a further development of the Hansen School, he has never based his work concretely on any scientific or explicitly ecological background. He approaches his designs with both a clear artistic intuition and sound horticultural knowledge. Nevertheless, the Hansen School's focus on the use of plants appropriate to a site is also ever-present in Piet's plantings. His knowledge is not drawn from predetermined lists, however, but from his long empirical experience as a perennial gardener. In this regard, his designs are indeed ecological and sustainable, as they respect the site and are created with a view to longevity. In terms of ecological processes and resilience, however, most of his concepts largely lack the element of dynamic development, self-renewal and spontaneity. His work must not be classified exclusively according to ecological or technical academic criteria - his plantings are to be seen as a form of artistic expression. If you ask Piet himself about his artistic aspirations, however, he will never claim to be an intellectual or an artist (see p.186); he is far too down-to-earth for that.

Piet's real achievement is in elevating the work of designing with plants to a whole new level. He has brought perennials back into the consciousness of landscape architecture. Renowned garden designers are now being valued increasingly for their specialist knowledge, and are even being brought in on high-profile projects and accepted as equal partners among architects. The

High Line in New York and the Lurie Garden in Chicago's Millennium Park are good examples of the new liaison between architecture and planting design. Even galleries and museums have come to value plants as a design tool for their outdoor areas (see pp.250–51) or even to include them directly in art projects, as at the Venice Biennale (pp.108–13). This has all helped to elevate an Oudolf planting design to a work of art, one that can stand on its own, as opposed to serving as an accessory or mere decoration. Piet Oudolf has thus succeeded in emancipating planting design as an art form in its own right, bringing it out of its niche and on to the big stage.

Piet, People and Plants

Jonny Bruce

What is remarkable about Piet Oudolf's success is how plants, and herbaceous perennials in particular, have always been given centre stage. So often, soft landscaping presents as a predictable supporting cast to the architectural lead, but not so with Piet. This was clear the first time I experienced his work, during a visit in 2011 to the Serpentine Gallery in London, where Swiss architect Peter Zumthor had been commissioned to create the annual pavilion. On entering the unassuming black box of a building, I was confronted by a fantastical meadow bathed in light from the large aperture above (see p.250), open to the sky and reminiscent of a work by the artist James Turrell. A low blue bench lined the space, inviting you to sit and contemplate the planting. An inversion of the traditional garden border backed against a wall, here the building becomes the frame. It was a *hortus conclusus*, celebrating the textural interplay of grasses such as *Molinia* 'Transparent' with robust perennials like *Eupatorium maculatum* 'Riesenschirm' – a manifest wildness characteristic of Piet's style.

To my twenty-one-year-old self, this collaboration was a revelation. Here was a garden centred in an art world context, and it helped adapt my prejudiced view of ornamental gardens as parochial and suburban. Around this time I decided that I would be a professional gardener, but it was another six years before I visited Piet and Anja's garden in Hummelo, in the Netherlands. This was the first of many visits, and Piet, generous with his time, would often lead me round the garden, peppering our walk with questions and observations about the plants. He would share his knowledge but also ask my opinion, as he noticed an unusual seedling, its flower bigger and brighter, its stem darker or its habit more upright than its neighbours', before disregarding it as 'not special enough'. It was through these walks that I developed an appreciation for his relationship to perennials – an innovative one seeking out the new and unknown but rooted in a tradition of growing, with multifarious influences and collaborators. It is Hummelo that continues to provide the most immediate and constant site for experimentation. As Piet explains, 'All the ideas grew in Hummelo.'

'It is just Nature now' – that is how Piet described the former nursery area of the garden at Hummelo to me on that first visit in the autumn of 2017. I had heard that he and Anja would be closing the garden to the public (this would happen the following year) and initially wasn't sure whether Piet's words held a certain resignation. Was it that he hadn't time for the garden, as he was so busy with international commissions? What I hadn't considered was that he was referring, somewhat provocatively, to the logical conclusion of an increasing but still contained wildness. In the large borders behind the main house, mare's tail runs riot, but somehow the planting feels anchored. Vertical accents of *Calamagrostis* × *acutiflora* 'Karl Foerster' shoot up at intervals, beating a rhythm among clouds of asters and expanding heads of *Eupatorium*. Everywhere there are points of interest. Recently exploded pods of *Asclepias* line the paths, their fluffy seeds catching on the sharp points of *Eryngium giganteum*. Behind, tall spires of biennial *Verbascum*, past their flowering but judiciously left standing, provide punctuation marks, while the increasingly scarlet canopy of sumac, held aloft on characterful multi-stems, adds dramatic colour to the scene.

In summer the garden is swathed in colour, and in winter the plants' skeletons provide a delicate grace, but it is for this autumn moment that Piet's style is most renowned. By weighting the perennial interest to later in the season, this approach ensures that the garden rarely looks tired at any time of year – a point eloquently made in Tom Piper's 2017 film, *Five Seasons: The Gardens of Piet Oudolf*. Bulbs and spring ephemerals are increasingly important to the designer's work, as can be seen from the project at Hauser & Wirth gallery in Somerset, UK, which opened in 2014, and his various collaborations with garden designer and bulb specialist Jacqueline van der Kloet. However, as Piet explains, earlier in the year you only need a few flowers here and there to hold interest, as the green rush of emerging perennials carries the scene. In *Designing with Plants*, the book Piet produced with Noel Kingsbury in 1999, he said that his 'biggest inspiration is Nature. I do not want

> I always had a strong feeling that I could do something different... and that happened when I met plants.
>
> Piet Oudolf

to copy it but recreate the emotion.' He is describing the emotion of walking through a wood and emerging on to an alpine meadow almost overwhelming in its diversity. More than anything, it is an attempt to evoke this balanced diversity of a wild meadow that underpins Piet's perennial plantings. Early on, he celebrated these plants through large-scale block plantings such as at the Dream Park in Enköping, Sweden – one of his earliest public commissions, from 1996 (p.249) – but over time he developed a more complex tapestry, where flowering plants are threaded through a matrix, usually of grasses. It is typical that what seems the most natural is the hardest to achieve, especially when combining plants that in nature would never grow together. Piet appreciates this, saying, 'the ultimate skill is to reach a certain harmony in all this variation.'

While Piet is rightly referred to as a pioneer of perennial design, his plant selection has changed remarkably little over the years. This is not to detract from his role as an innovator, but rather to highlight how much of his contribution is grounded in recognizing the value of old varieties that had become neglected. As Piet reminds me, 'a good plant never goes out of fashion.' Many of these older varieties were developed in specialist nurseries in Germany by the likes of Georg Arends (1863–1952), Ernst Pagels (1913–2007) and Karl Foerster (1874–1970); plants like *Miscanthus sinensis* 'Flamingo' and 'Kleine Fontäne', grasses selected by Pagels, whose attractive colour and habit have stood the test of time. It was also Foerster who wrote as early as 1957 of grasses – a plant synonymous with the Oudolf brand – 'How can these jewels of the garden have been ignored for so long?'

Piet readily admits the influence of others in both his plant palette and his approach to design. Of particular early significance was the Dutch landscape designer Mien Ruys (1904–1999). Ruys dominated urban landscaping in the post-war Netherlands, underpinning her practice with a democratic ideology of accessibility to green space. Her clean, modernist hardscape provided a simple foil to the intelligent abundance of her planting. She was a designer who put plants first, and although Piet's work quickly developed its own character, in a sense she gave him permission to do the same.

In the 1970s and 80s there were very few perennials available in the trade, and it became clear to Piet that he would have to grow his own. In 1982 Anja and Piet and their young family bought a derelict farmhouse with just over one hectare (three acres) in Hummelo, about thirty minutes from the German border. Piet explained, 'It was not really about living in the countryside, but having space to grow plants.' On arrival he set about establishing a nursery.

At first Piet partnered with Romke van der Kaa, who had been head gardener at Great Dixter in Sussex, England, but from 1985 it was increasingly Anja who ran the administration, while Piet focused on propagation. They were always looking for unusual plants and sourced seeds from botanical gardens around the world. The nursery was intended to supply perennials for Piet's design projects, but it quickly took on a life of its own. In 1983 the first open days were held, when nurserymen came from across Europe to sell plants. These annual open days, which came to an end in 1997, became a bustling exchange not only of plants but also of ideas. In that first year Piet and Anja invited Rob Leopold, a uniquely cerebral plantsman who had studied philosophy at Leiden before setting up his own native seed company, Cruydt-Hoeck. Leopold was passionate about conserving wild plant communities at risk from the rapid industrialization of the Dutch countryside. However, Piet most valued Leopold as a connector. As he explained, 'Rob's ultimate strength lay in his ability to play the role of catalyst, to encourage people to discuss and think about their activities, and to find new opportunities.'

It was this stimulating exchange of ideas, coupled with its convenient geographic position between the United Kingdom and Germany, that made Hummelo a horticultural hub. Piet and Anja also travelled. Piet explains, 'We would go every year to visit the gardens and nurseries of John Coke, Elizabeth Strangman, Beth Chatto and Alan Bloom. These were the people with the best plants.' As their network expanded, so did the number of trial beds at Hummelo. They became aware that there were certain genera whose cultivated potential had not been realized. To this end they rented 4,000 square metres (just under an acre) of land to grow thousands of seedlings of plant species such as *Astrantia*, *Echinacea*, *Salvia*, *Monarda* and *Sanguisorba*. Some plants, such as the long-flowering *Gaura* 'Whirling Butterflies', were chance encounters, but others took years of careful selection.

Two notable selections among the many plants that grew out of this process were *Astrantia* 'Roma' and 'Claret'. 'Roma' is a robust and long-flowering pink colour (see p.263), while 'Claret' is deep wine-red. It took more than six years of selecting to produce 'Claret' as a stable seed strain in which germination would reliably produce a deep red flower. In contrast, 'Roma' is sterile and therefore must be propagated vegetatively. Nevertheless, it is this sterility that causes such impressive flowering. After pollination, many plants will reduce flower production to divert resources into the developing seeds, but with no viable seeds, 'Roma' continues to turn out blooms with abandon. This process of trialling provided an uncommon knowledge that continues to set Piet apart from other designers. As James Corner – who led the design for the High Line in New York City – described him, 'Piet's like a really good chef … he knows his ingredients.'

The primary inspiration for this systematic trialling and selecting of plants was Ernst Pagels, described by Piet as 'my hero for his unconventional way of looking at plants'. A student of Karl Foerster, Pagels had a huge influence on this new generation of Dutch growers. It is testament to Pagels' uncanny ability at selecting plants that so many of his 130 or so introductions are still in cultivation. Plants like *Veronicastrum* 'Lavendelturm', with its graceful spires of pale lavender, the verticals of *Phlomis tuberosa* 'Amazone' and the deep blue of *Salvia nemorosa* 'Ostfriesland' are just a few of Pagels' plants that continue to play key roles in Piet's plantings. In the Lurie Garden in Chicago, which opened in 2004, Piet experimented with a broader palette of American native species but still retained these old favourites. As with his early Dream Park in Sweden, the Lurie Garden also features an iconic salvia river, where broad ribbons of *Salvia nemorosa* cultivars run through the plantings (see p.80). In early summer these transform into swathes of purple and blue as the different varieties, including Pagels' 'Ostfriesland', blend delicately together.

In order to protect their selections, Piet and Anja became partners in the plant distribution company Future Plants, which helped produce the plants at the scale needed for Piet's increasing international commissions. The company also helped connect the designer to growers, particularly in America, including native plant enthusiast Roy Diblik of Northwind Perennial Farm in Burlington, Wisconsin, who consulted on the Lurie Garden. Of great significance was a joint trip to the Schulenburg Prairie in 2002, now more than 40 hectares (100 acres) of restored landscape managed by the Morton Arboretum in Lisle, Illinois. Piet admits that seeing this vast grassland in early summer, punctuated with the pale candles of thousands of *Baptisia*, was 'a real turning point'. He would return many times in different seasons – including late summer, when golden *Andropogon*, *Rudbeckia* and *Silphium* begin to push through a sea of pink *Echinacea*, while pale umbels of *Parthenium* and thistles of *Eryngium yuccifolium* mix with the flowering *Sporobolus*. 'Who cannot fall in love with that?' exclaimed Piet incredulously. It was after these visits that he returned to the Lurie plans, incorporating more matrix-style plantings of native species to evoke the wild prairie (see pp.78–81).

Piet's contagious enthusiasm reflects his experience and the generosity he has received, and is a legacy as important as the plants themselves. Certainly, walking with Piet in his garden is never a passive experience: as he runs his hands through the grasses overhanging the paths, he directs your attention to some uniquely beautiful seedhead before raising his phone to capture it. It is a way of seeing that developed in those early days at Hummelo, embracing wildness and appreciating the beauty at every stage of a plant's life.

The person who helped form his philosophy, and whose eyes Piet especially valued, was the ecologist, activist and artist-gardener Henk Gerritsen. Piet admits that Gerritsen, with his deep knowledge of native plant communities, helped 'guide the planting into a natural setting'. The garden at Priona, in Schuinesloot in the Netherlands, which Gerritsen created with his partner, Anton Schlepers, pushed the envelope of wildness further than anything Piet had yet attempted. Here, aggressive 'weeds' such as ground elder were encouraged into competitive communities which, though experimental, were not always the most visually interesting. 'For Henk it was more about philosophy', says Piet, but considering the mare's tail at Hummelo, he wryly admits that Gerritsen probably would approve. The back-and-forth of plants and ideas between Priona and Hummelo found expression in their first book, *Droomplanten* ('Dreamplants'), published in 1990; it was published in English in 2003 and again in 2019 as *Planting the Natural Garden*. It was with Gerritsen that Piet learned to appreciate plants outside their flowering season, even in a state of decay. Both Gerritsen and Schlepers would die from AIDS-related illness, and Gerritsen wrote movingly how 'people used to be so afraid of death in the garden. Every yellow leaf was an imperfection ... but now a whole generation has known death, we do not ban it anymore.'

For Piet, the aesthetic value of decay is primary, but there is also an intimacy about the garden at Hummelo, where so many plants tell a story of a cutting taken, but more often given. *Stachys* 'Hummelo', for example, a strong, floriferous form of betony with deep purple-pink flowers, is deservedly popular and often attributed to Piet – though it was actually a selection by Ernst Pagels. Pagels gave it the name 'Hummelo' as a gift to Piet and Anja, but in fact it refers to the plant's qualities as a pollinator plant – from the German *Hummel*, meaning 'bumblebee'.

Hummelo still remains a place of experimentation. In recent years Piet has been excited by developments with *Vernonia* (see p.278). Most of these North American ironweeds are much too large for normal garden use, with coarse foliage, but recent crosses using narrow-leaved species like *V. angustifolia* and *V. lettermannii* have produced more garden-friendly plants, such as *V.* 'Summer Swansong' and *V.* 'Iron Butterfly'. He has also been experimenting with many of the new hybrid *Baptisia*s. With excellent drought-tolerance for use in hot, urban plantings, and a kaleidoscopic range of colours, these plants have found deserved popularity – though Piet remains fond of the wild species, finding that 'many of the new selections are too unnatural'. Grasses are still a mainstay of the garden, but introductions such as *Molinia* 'Dark Beauty', with beautifully dark stems in autumn, and *Schizachyrium* 'Ha-Ha Tonka', an improved compact form of the flop-prone little bluestem, show that he has not lost his appetite for novelty.

'There are still so many plants to discover', Piet says, with characteristic enthusiasm. It was always a pleasant surprise on taking Piet and Anja a box of plants to hear the words, 'I haven't grown that before', and to think that our conversations were finding their place alongside those of Henk Gerritsen, Roy Diblik, Rob Leopold and the countless other gardeners who have informed this rare garden. It is in this way that one can best understand Piet Oudolf, for as he said, 'My relationship to plants is through people.'

I'm not a painter,
but I can see things in my mind
that I can translate
into the ground.
I can see things
that will happen over time.

Piet Oudolf

THE DRAWINGS & PLANTING PRINCIPLES

Perennial Perspectives

Noel Kingsbury

The first sight of a Piet Oudolf garden in late summer reveals an expanse of softly textured plants, not colourful in the way that perennial plantings have traditionally been, but a dense tapestry that seems to appeal to multiple levels of sensory appreciation. The vast majority of people who know Piet's work are familiar with it from arrays of massed perennials and grasses. To create these plantings, he sketches and draws his designs using plans that until now have been seen only by a minority of his admirers. So striking are these plans - done with coloured pencils, felt-tip pens and markers, and using a dense language of annotation, including dots, dashes, spirals and squiggles - that they have become accepted as artworks in their own right, with several galleries holding exhibitions of them (see also p.24). The plans are, in a way, representative of the creative process of an artist - a process that is next to impossible to explain, in part because the routes of the design method are many and may vary greatly from one project to another.

The planting plan occupies an odd position in landscape design. Regarded as almost essential to win over the interest of clients, it can be interpreted only by the few who are familiar with all the plants. What plans are good at is conveying the diversity of the plant material used and a sense of its spatial distribution; close reading enables plants to be located in a list and then looked for on the accompanying drawing, while the distinctive icons used to illustrate each species or variety allow the viewer to identify them.

So what do we see when we view a Piet Oudolf planting? Why do we like what we see? Answers to questions like this inevitably involve theories of aesthetics and, even more fundamentally, theories of visual perception. But is our response instinctual or culturally determined? The belief that our responses to visual cues are somehow hard-wired has been around for a long time. Consider, for example, the Golden Rectangle and Golden Ratio discovered by the ancient Greeks and much used by artists ever since, and the prospect-refuge theory often discussed by landscape designers, which seeks to explain our appreciation of topography in terms of our evolution in the African savannah, leading us to favour relatively open country with a scattering of trees and small areas of woodland.

'Organic' responses to these plantings - that is, reactions derived from innate, genetically determined responses to visual cues - may be sought in such theories of visual perception as Gestalt theory, models that seek to explain how we interpret what we see in terms of a series of laws governing the hard-wiring of the brain. The repetition of plants, for example, enables our minds (usually subconsciously) to read wherever they are growing as a group, therefore unifying and defining a particular area. Some favourite Oudolf plants are very obvious rhythm markers, much as the beat of a drum is in music: tall grasses, such as *Calamagrostis × acutiflora* 'Karl Foerster', or distinct flower spikes like those of *Agastache, Baptisia* or *Veronicastrum* species (see pp.18-21 and 132-33). But everything in Piet's planting repeats, so the levels of perception of unity-building visual reinforcement must go very deep and should be regarded as a key part of the visual impact of his work.

At the same time, however, our perceptions of Oudolf plantings are, of course, strongly culturally conditioned. There is no doubt that we respond to Piet's plantings at this moment in history in a way that would have been inconceivable if someone had started laying out large expanses of grasses and perennials in the 1920s, let alone the 1820s. We live at a time when we want to appreciate nature, and are far more ready now than 50 years ago to view plantings with no straight lines or the other traditional elements of human organization, once regarded as so necessary that their presence was what helped define a garden.

Two other aspects of Oudolf plantings also make us respond: the concentration on plant structure and tonal depth, both of which rely on a clearly articulated language of plant form: the spikes and spires, umbels, daisies, plumes, buttons and curtains discussed in our 1999 book, *Designing with Plants*. The book seeks to group plants roughly, based on common shapes, using

I discover beauty in things that on first sight are not beautiful. It is the journey in life to find out what real beauty is and notice it is everywhere.

Piet Oudolf

these as a guide to selecting then grouping plants for visual effects that are stimulating but at the same time harmonious.

Piet's plantings create a landscape of forms and textures that, beyond a few obvious standouts, is not easily consciously read. Our visual sense tends to get distracted by colours, and though form, texture, seasonal value and colour are all elements of his designs, Piet insists that he is not a colour gardener. This may or may not be disingenuous; flip a photograph of an Oudolf planting into black and white, and it retains enormous visual interest and enough clarity for us to make out virtually every species used. Allied to this is the distribution of tonal depth, the range between the dark (visually receding) and the light (visually prominent). Tonal depth is most clearly appreciated in late autumn and early winter, when pale grasses and dark seedheads contrast highly effectively (see pp.147, 156–57). An irony here is that the visually bright grasses are often formless, almost inchoate masses of seedheads, whereas the dark shapes are usually sharply defined seedheads or dead stems. The contrast between bright/formless and dark/defined is very powerful. However, earlier in the year – in spring and early summer – this powerful aesthetic is not as available, so plant form and colour inevitably become more important (see p.146 top, p.225 bottom).

The focus on plant form should perhaps be sought in Piet's early career. Starting work as a garden designer in the 1980s, he was inevitably influenced by the Dutch landscape and garden designer Mien Ruys (1904–1999), with her pioneering designs using perennials (see also p.33). Ruys combined woody plants and perennials, using the former very often geometrically clipped, a traditional practice to which she brought a post-Bauhaus modernist eye. Much of Piet's early work featured clipped foliage, often used in highly innovative ways. By the late 1990s he was abandoning this approach, but arguably transferred the ability to play with graphic forms from clipped shrubs to perennials. Woody plants he used from now on would be 'grown free'. In a reversal of the Arts and Crafts style that dominated twentieth-century British practice, whereby a framework of clipped shrubs contained more unruly perennials, the Oudolf style uses the more amorphous forms of naturally grown woody plants to frame and provide a counterpoint to the perennials that provide the primary graphic interest (see pp.150–51, 154–55, 199).

Piet's own skilful photography of his work has enabled his vision of what he does to dominate our perception of it, through books that appeared from 2000 onwards. Published imagery of naturalistic plantings by other designers only serves to reinforce the trend, but the clear visual articulation of Piet's plantings through juxtapositions of form and texture always makes his work the most memorable. The more genuinely naturalistic, or ecologically dynamic, intermingled plantings favoured by other practitioners suffer, perhaps, from an inability of many non-gardeners or non-ecologists to read and appreciate what is in front of them.

One of the secrets of the spatial arrangement of plants in an Oudolf planting, described as block-matrix-scatter, is Piet's ability to produce plantings that are highly coherent, in the sense that they are readily interpreted by the uninitiated but also sophisticated enough to appeal to connoisseurs of horticultural design. The drawings offer another way in to appreciate his work. Somewhat ironically, these are far more highly-coloured than the result they portray, the colour and range of markings emphatically illustrating the complexity of the spatial relationships of the plants used. In many ways, Oudolf has not left a traditional trope of conventional planting design, that of grouping individuals of a given cultivar or species in a block, an approach clearly articulated by Gertrude Jekyll in the 1920s. Rather, he has developed nuanced effects that give it a new lease of life, such as the use of 'scatter plants' – loose groups of individuals of species that are prominent for a defined period (usually because they are in flower), but which are relatively inconspicuous before or afterwards. Nearly all his plantings from the Lurie Garden (completed 2004) onwards use this technique to some extent (see pp.78–9), including the garden at Hauser & Wirth in Somerset and Belle Isle in Detroit (see p.130, 226).

Species that are conspicuous after flowering and look a mess are used only occasionally and cautiously, sometimes allowed in for a valued early flowering. In the 1990s Piet would refer to these as 'filler plants', but he rarely mentions them now. The overwhelming bulk of his plants have good visual structure for a relatively long time. It is one of his most important design principles that plants have to look reasonably attractive after their peak flowering; indeed, it can be safely said that every Oudolf planting looks almost as striking after its peak flowering as it does during it, and often more so than before (see p.156–57). It is possible to analyse the frequency of distribution of plant cultivars in an Oudolf planting and group them: the most frequently occurring tend to be those with the longest season of strong appearance (structure and texture rather than colour, e.g. *Veronicastrum virginicum*, Culver's root); those used with reduced frequency are either slight variations of those in the former group, or those that tend to have a shorter season of visual appeal. There is also a third group: cultivars that appear at low frequency. Oudolf designs for a large public planting may well include over 100 taxa (species and cultivars), and occasionally more than 150, which is far more than most designers would use, wilfully challenging the 'less is more' doctrine. And yet it doesn't. These 'low frequency' species are a very small part of the whole biomass, but they add surprises, they keep up the interest in walking around the paths that typically cut through Piet's planting designs. The analogy might be with rare or infrequent plants in a wildflower meadow or other habitat; they are statistically insignificant but add to visual and bio-diversity.

Planting design involves the creation of scenes that must be appreciated at multiple, very different levels, which is perhaps one of its greatest challenges. The macro-view must impress at first sight, when the eye is cast over several thousand square metres, but there must also be interest at closer range, as one walks through; crucially, there must be interest at the micro-level, within a few metres, as one all but brushes past the plants. 'Frequency' and 'balance' are the buzzwords that Piet uses here. He stresses that plants have to look good next to each other, but also that the totality has to be beautiful. The drawings and plans, once the onlooker has got their eye in, enable an appreciation of this; juxtapositions can be spotted easily, spatial distribution also seen and evaluated.

The commitment to planting in monocultural blocks, which has always lain at the heart of his work, may be subject to many subtle techniques to break it up, but it does set Piet apart from what has become the zeitgeist of the burgeoning field of twenty-first-century planting with perennials. Led by a serious body of research conducted at a number of German institutions (mostly technical universities) from the 1990s onwards, 'intermingling' is the key word. In this system, plants are not, or only rarely, grouped in blocks; instead, several species or cultivars (often up to 15) are grown together, more or less randomly combined, the intention being to create a kind of mock ecosystem; the results can be seen in many of the impressive German parks that are the legacy of garden shows. The totality, the macro-view, is granted more importance than the details of exact juxtaposition. For Piet, this is an abandonment of a key principle. 'Why do you design? Because you want to have control', he says. A looser style of planting can be seen in the 'matrix' combinations he has been using since the late 1990s. Here, a mass of a visually retiring, ground-covering species, usually a grass, is dotted with visually stronger plants. The overall effect has some practical considerations (potentially reducing maintenance), looks more naturalistic and makes an effective contrast with the block planting (see pp.100–01, 170).

The development of a number of distinct genres of planting has also driven the transformation of Piet's planting plans. The early ones, in conventional black and white, with blocks indicating multiples of particular taxa (see p.53), underwent considerable development in order to accommodate the introduction of scatter plants and then matrix planting (see p.62). Block planting and its variations is what makes Piet's plans function as artworks in their own right. The expression of an intermingled planting in his style of drawing would more resemble a pointillist painting and would only be realistically feasible to produce if computer-generated. Piet's planting has tended to be highly seasonal, which has also been one of the main criticisms levelled at it. Increasingly, though, he has started to use bulbs (see p.105, 230) to provide spring colour, and his use of woody plants (pp.64, 184) has been far more extensive over a longer period than many of his fans realize. The visual impact of his work with these plants is very different from that of other designers, however, and much more subdued. Bulbs in spring offer very little of the structure and texture that dominates our perception of the summer-to-winter perennial growth. They do offer colour in his plantings, but much more subtly than the conventional use of tulips and larger daffodils varieties do. Natural species, or derivatives of them, that maintain a size and almost identical proportion to their wild ancestors, are favoured. This, together with the use of grasses, accounts for our reading of these plantings as being 'naturalistic'.

The palette of broad-leaved (non-coniferous) trees and shrubs available in temperate regions is relatively uniform visually; shapes tend to be similar (and often rather amorphous), with leaves that are not large enough to stand out at a distance. There is nothing like the range of form and texture to enable the kind of visual articulation that makes Piet's perennial plantings so distinctive. And because they take much longer to mature, the realization of the designer's vision is one that only future viewers will appreciate. Autumn colour, spring blossom, attractive bark colours and textures all contribute interest, but usually for clearly defined periods. Specialists have plenty to get excited about, but for most viewers, the woody plants in an Oudolf planting are appreciated only on a 'macro' level: framing, breaking up and directing vistas, and providing context for what will always be the main course of the feast - the perennials. Nevertheless, given plant lifespans and the vicissitudes of maintenance regimes, it may well be that this part of his work - the woody plants - will best stand the test of time.

Inevitably, given the transitory nature of gardens in general, and of perennial-based planting schemes in particular, these places will change over time, often dramatically. The plans at least will preserve their originator's thoughts and intentions for posterity. Whether they can be seen as preserving the design process behind them is another matter - and one which would be dependent on someone having developed an appropriate algorithm to analyse them. It would be nice to think that this may prove an impossibility, and the plans will serve future generations as inspiration and guide rather than a text or recipe to be followed.

Looking at Piet's planting in the broad sweep of history, it can perhaps be best understood as representing a stylization of nature, much as the highly influential landscape movement of eighteenth-century England created a highly structured representation of the natural world. Much as it went on to stimulate a very wide range of design styles, we can speculate that Oudolf planting will also have a major and continuing influence on the nature-inspired planting of the future.

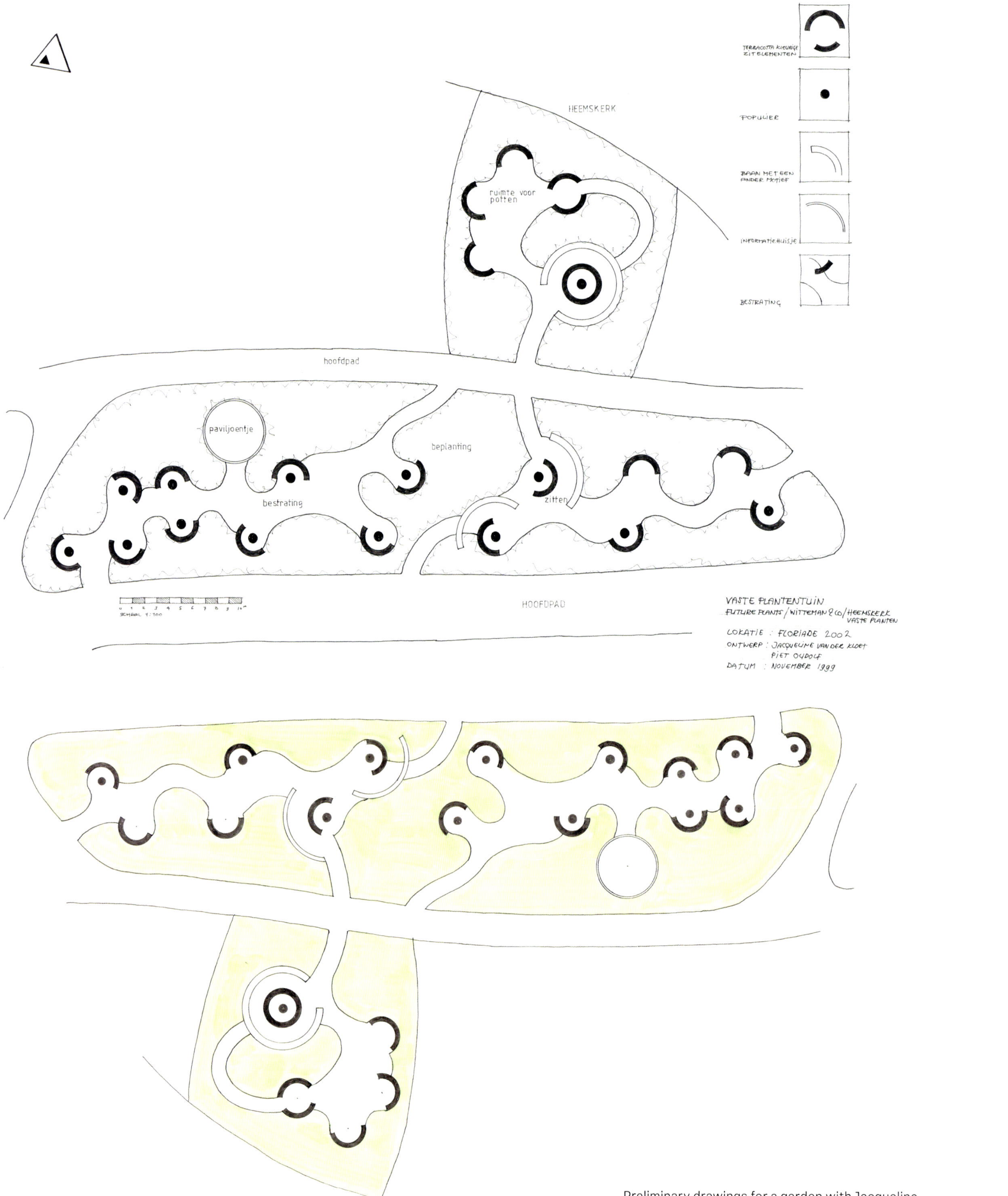

Preliminary drawings for a garden with Jacqueline van der Kloet for Future Plants (a company promoting new plant varieties, of which Piet used to be a partner) at the Floriade in 2002, a major Dutch garden festival held every ten years. Note the very simple, spare drawing style.

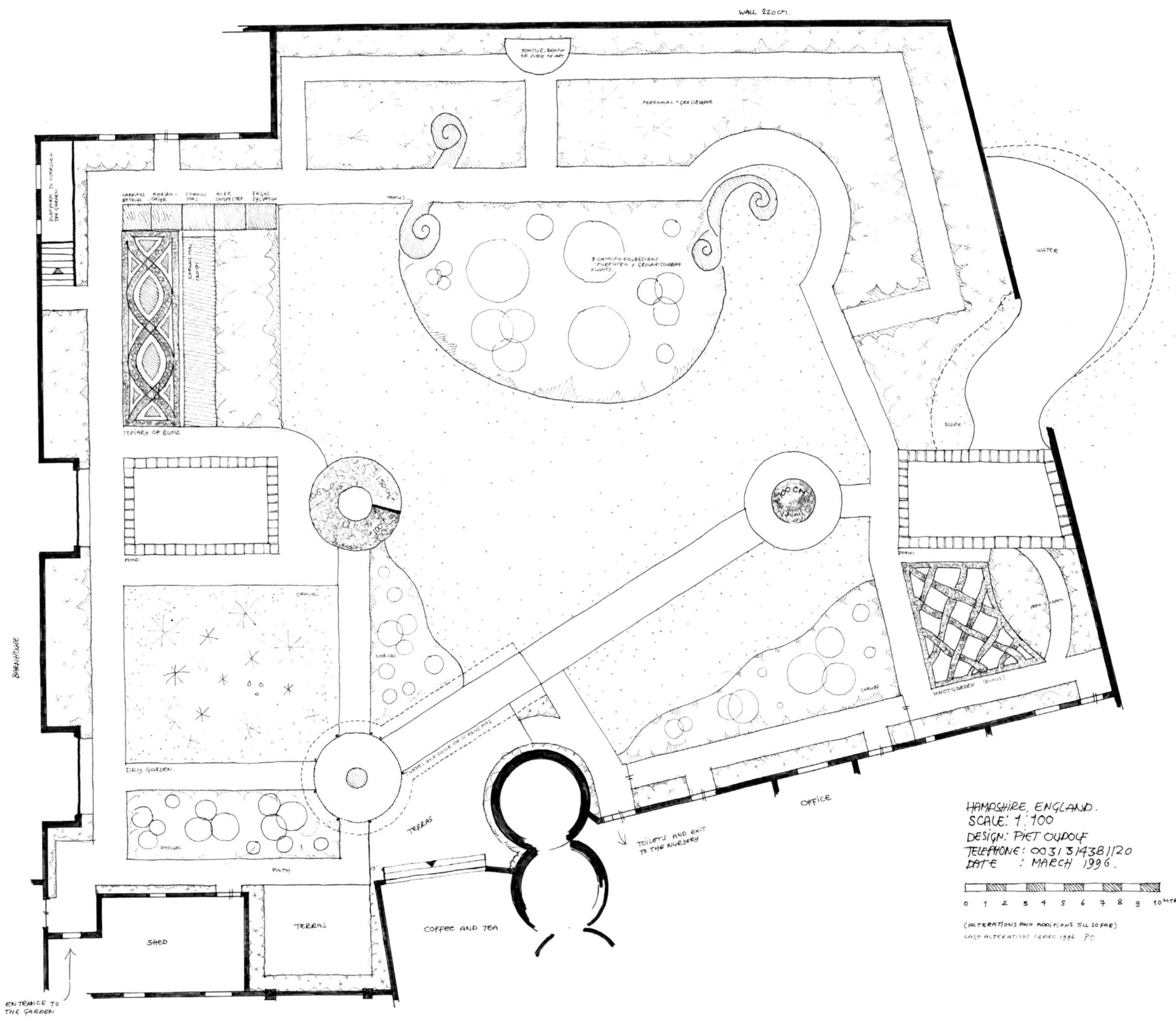

Plan for a courtyard garden at Bury Court, Surrey, for the plantsman John Coke, 1996. This was Piet's first project in Britain. Like all early plans, it is simply ink on paper and at this scale, no individual perennials are indicated. The geometrical elements are typical of his style at this point, something that may surprise those who came to his work later. The large semi-circular area at the top became the 'Deschampsia meadow', a prototype for the matrix plantings of later years.

Section X.1
July 3rd 2008
Nantucket
0 2 4 6 8 10M
1 100 metric
REVISED NOV 25 - 2008

Trees and Shrubs

AC	Acer campestre
AG	Acer ginnala
AT	Acer truncatum
AL	Albizia julibrissin
BL	Black Locust
CO	Celtis occidentalis
CK	Cornus kousa
HA	Halesia tetraptera
HL	Honey Locust
MsHL	Honey Locust multistemmed
MA	Magnolia hybrid
MAv	Magnolia virginiana
NYS	Nyssa sylvatica
PAR	Parrotia persica
PS	Prunus autumnalis
PY	Prunus yedoensis
QA	Quercus alba
QI	Quercus ilicifolia
QP	Quercus palustris
QGP	Quercus palustris 'Green Pillar'
QR	Quercus robur
S	Sassafras albidium
✱	Scotch Pine
STY	Styrax japonica
TAX	Taxodium ascendens
AE	Aesculus parviflora
AM	Amorpha fruticosa
ARM	Aronia melanocarpa/arbutif.
BE	Betula nigra 'Heritage'
BE Lk	Betula nigra 'Little King'
BE Ms	Betula multistemmed
cdi	Callicarpa dichotoma
chm	Chaenomeles 'Moerloosei'
chn	Chaenomeles nivalis
CR	Chionanthus retusus
CV	Chionanthus virginicus
CLER	Clerodendrum trichotomum
CL	Clethra alnifolia
⊙	Cornus mas
c.vir	Cornus sang. 'Viridissima'
cwf	Cornus ser. 'Winterflame'
COT	Cotinus coggygria
CRC	Crataegus crus-galli
CRP	Crataegus phaenopyrum
CRW	Crataegus 'Winter King'
▲	Evonymus atropurpureus
FM	Fothergilla major
+	Hamamelis vernalis
⊕H	Hydrangea heteromalla
⊕M	Hydrangea mariesii
⊕V	Hydrangea veitchii
H	Hydrangea villosa
●	Ilex 'Dr. Kassab' / Ilex opaca
IG	Ilex glabra (Inkberry)
IL	Ilex verticillata (Sparkleberry)
✕	Juniperus virginiana
O	Ligustrum ovalifolium
MC	Maackia amurensis
b	Myrica pensylvanica (Bayberry)
PH	Philadelphus coronarius
PT	Populus tremuloides
PM	Prunus maritima
P	Prunus virginiana
RHA	Rhus aromatica
RHY A	Rhus typhina size A
RHY B	Rhus typhina size B
RH	Robinia hispida
rc	Rosa carolina
rg	Rosa glauca
roa	Rosa 'American Pillar'
rob	Rosa 'Blush Rambler'
rof	Rosa 'Francis Lester'
rom	Rosa moyesii geranium
ros	Rosa sericea pteracantha
row	Rosa 'Wedding Day'
rk	Rosa 'Kathleen Harrop'
rr	Rosa rugosa white/pink
Schy	Schyzophragma hydrangeoides
@	Shepherdia argentea
SP	Salix purpurea
SR	Sambucus racemosa
STE ps.ca.	Stewartia pseudocamellia
ST	Stewartia monadelpha
syp	Syringa meyeri palibin
syl	Syringa persica laciniata
T	Tamarix ramosissima
VAC	Vaccinium corymbosum
✱	Viburnum dentatum (Arrowood)
VD	Viburnum dilatatum 'Asian Beauty'
VF	Viburnum farreri
VN	Viburnum nudum 'Winterthur'
VA	Viburnum 'Allegheny'
VO	Viburnum trilobum
V	Vitex agnus-castus
KL	Kalmia latifolia
CF	Calycanthus floribus

Plan and legend for detailed planting for a woody plant screening of a large private garden on Nantucket island in Massachusetts, 2008: wind from the ocean is a major problem here.

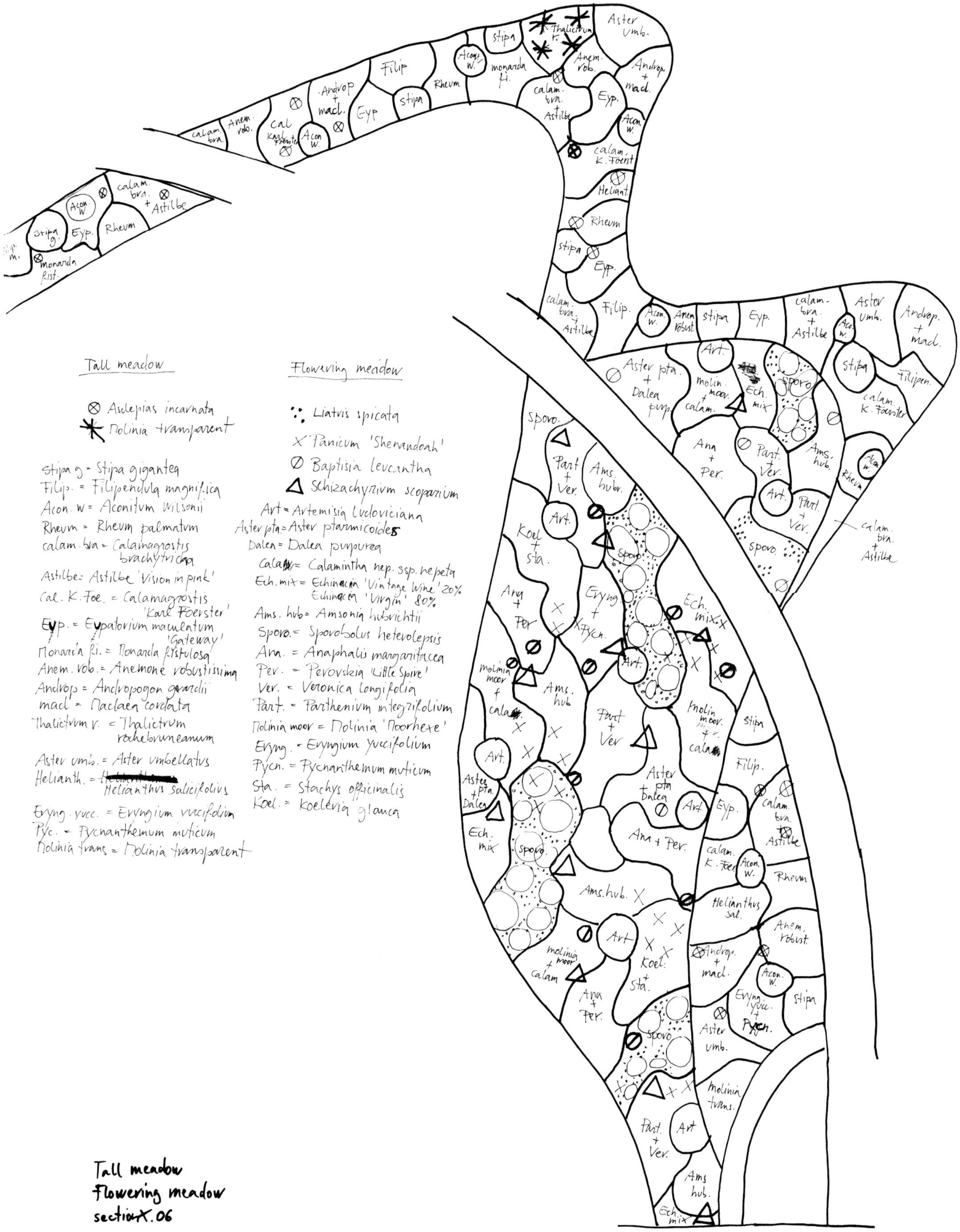

Sketch plan for a planting in the garden on Nantucket island, Massachusetts, involving perennial borders of two different heights, 2008. Plants are indicated by abbreviations (as is most common with designers in the English-speaking world) and also some symbols (more common among German practitioners).

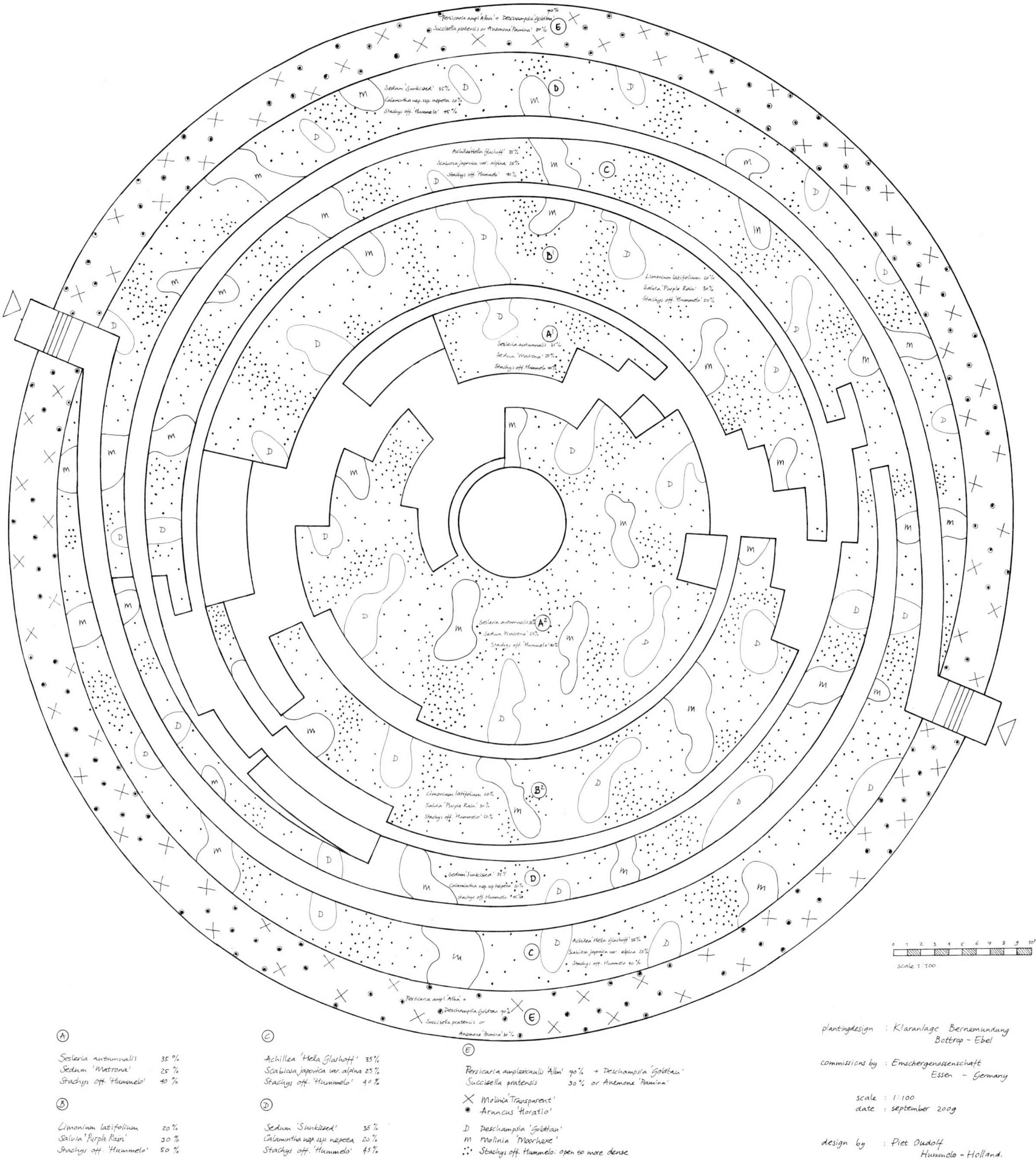

Plan for a perennial planting and walk-through garden, entitled Theatre of Plants, in a filled-in wastewater tank, as part of a post-industrial redevelopment project with Gross Max landscape architects at Bernepark, Bottrop-Ebel, Nordrhein Westfalen, Germany, 2009.

Overleaf: Sketch for a shaded area for a public park in Skärholmen, Stockholm, which was planted in 2011. The arrangement of the plants reflects the more extensive clump-forming habit of woodland species.

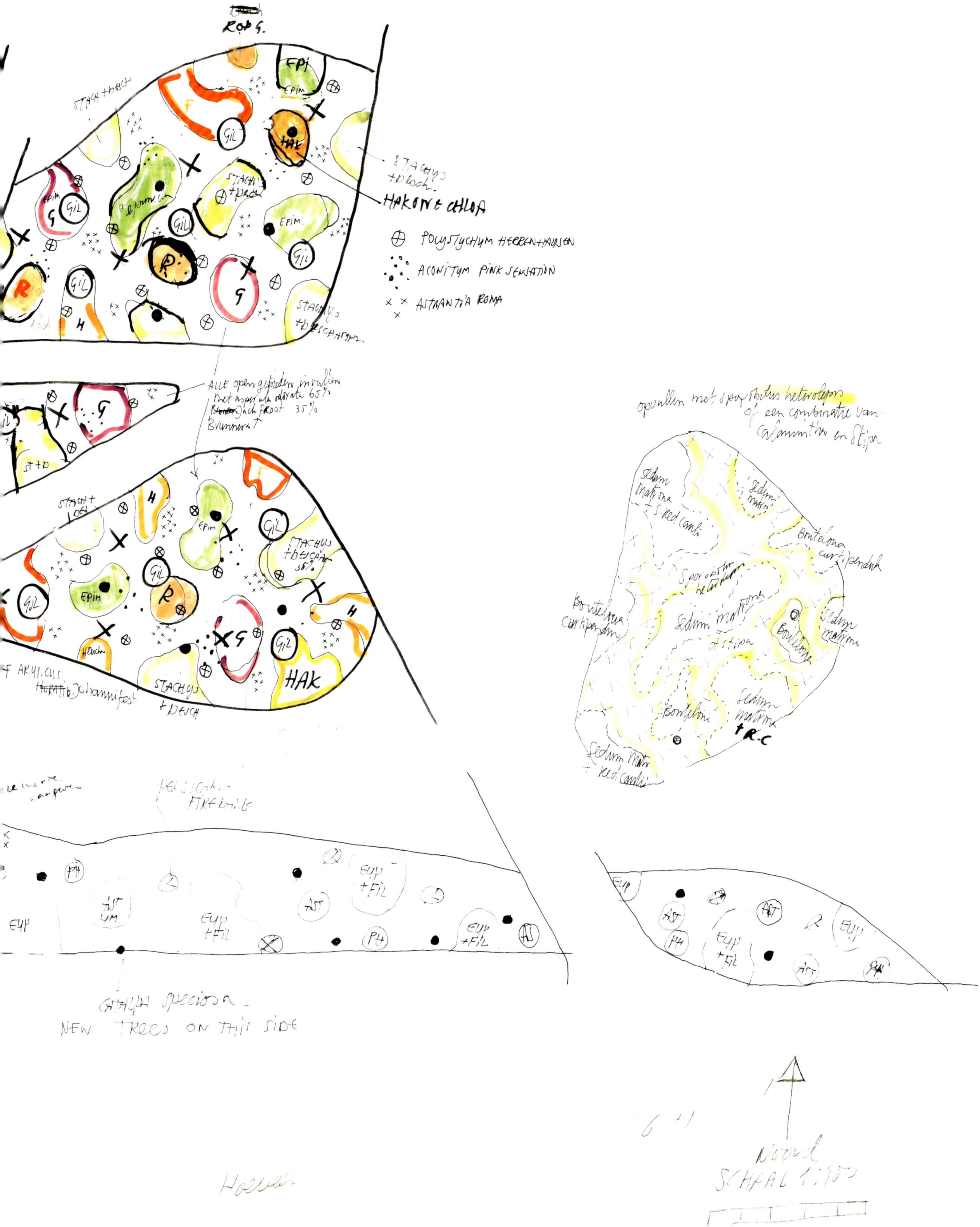

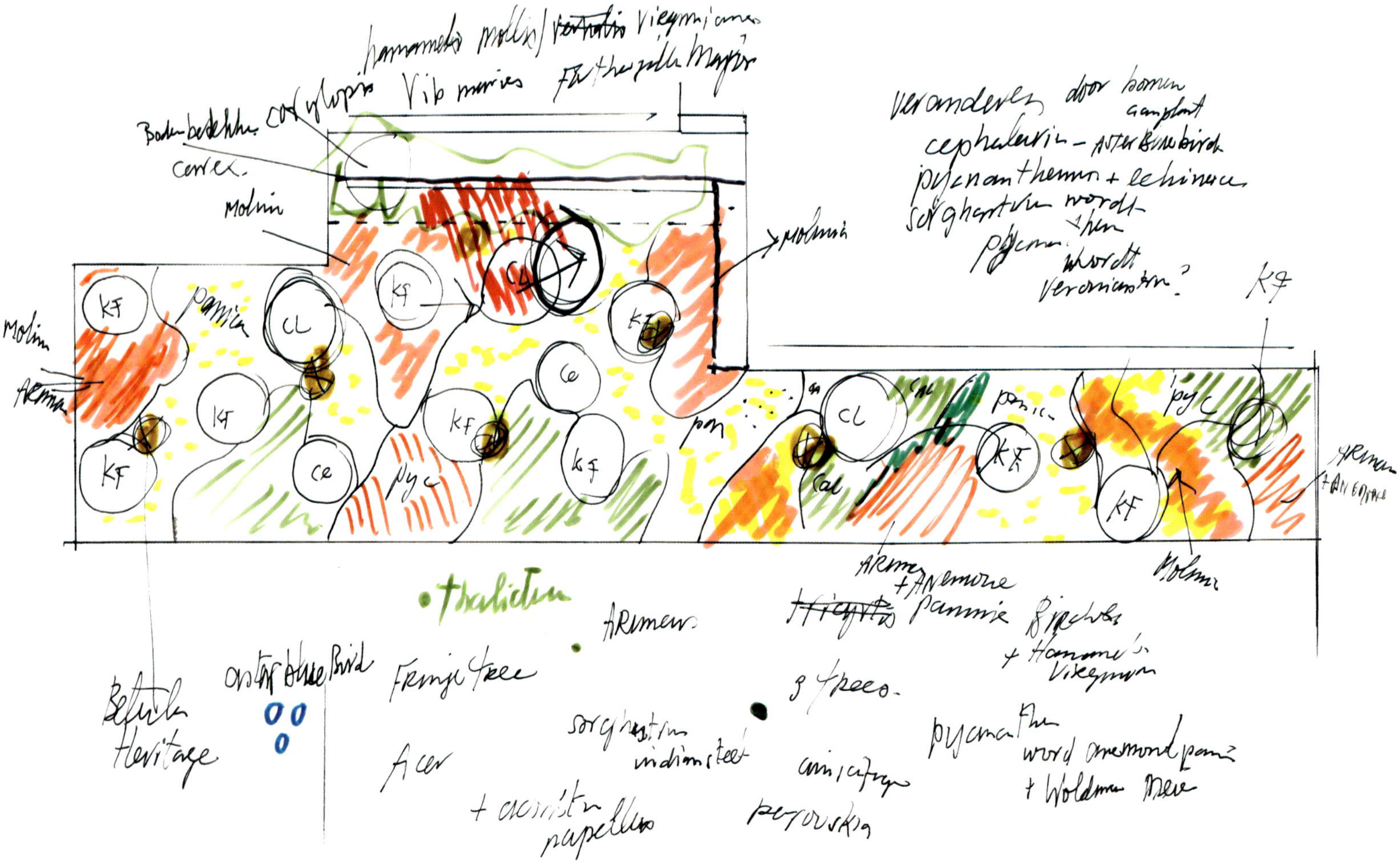

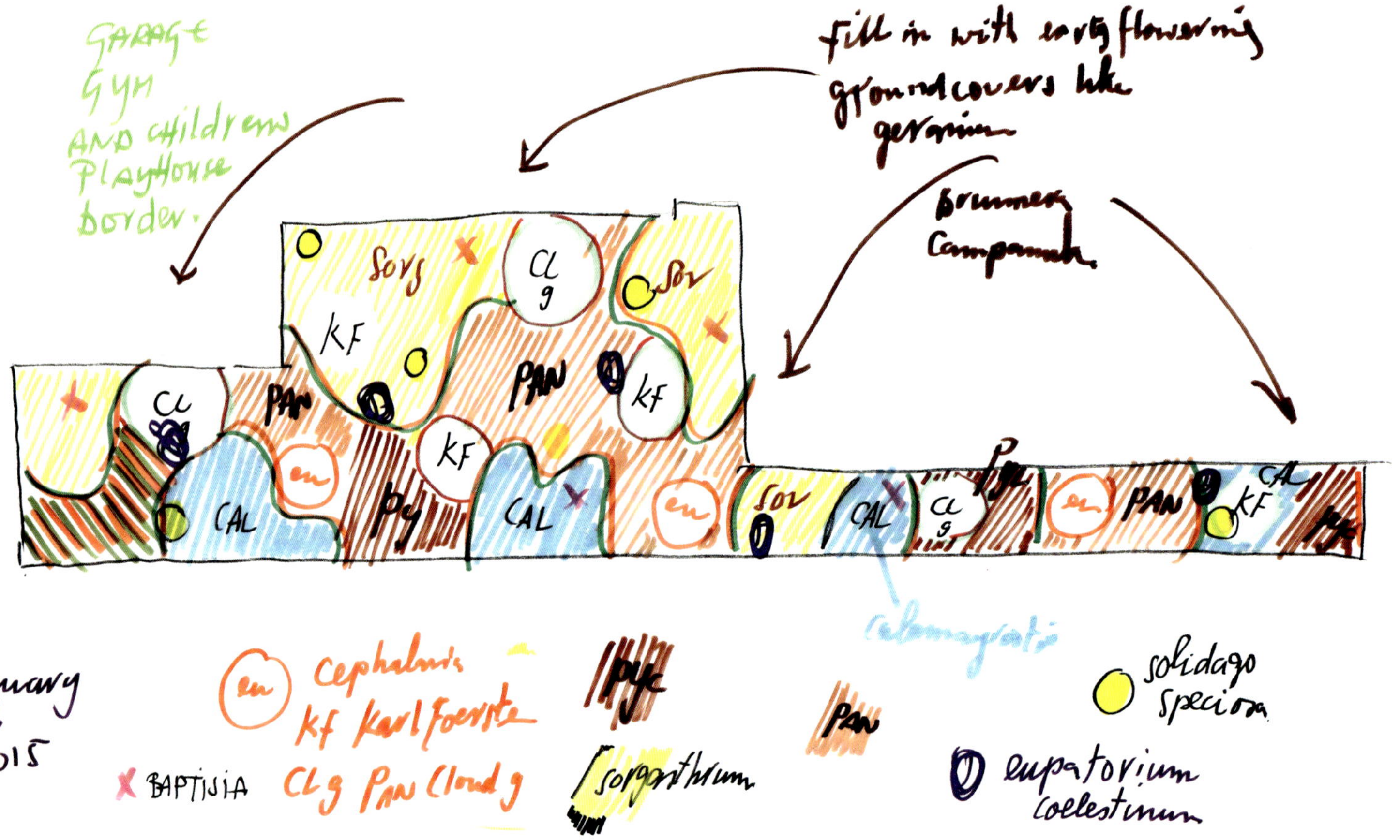

Sketches (above) for planting in front of a house with birches and (right) for a temporary planting at a garden show in Nagasaki, Japan.

These working drawings should be seen very much as 'thinking aloud'.

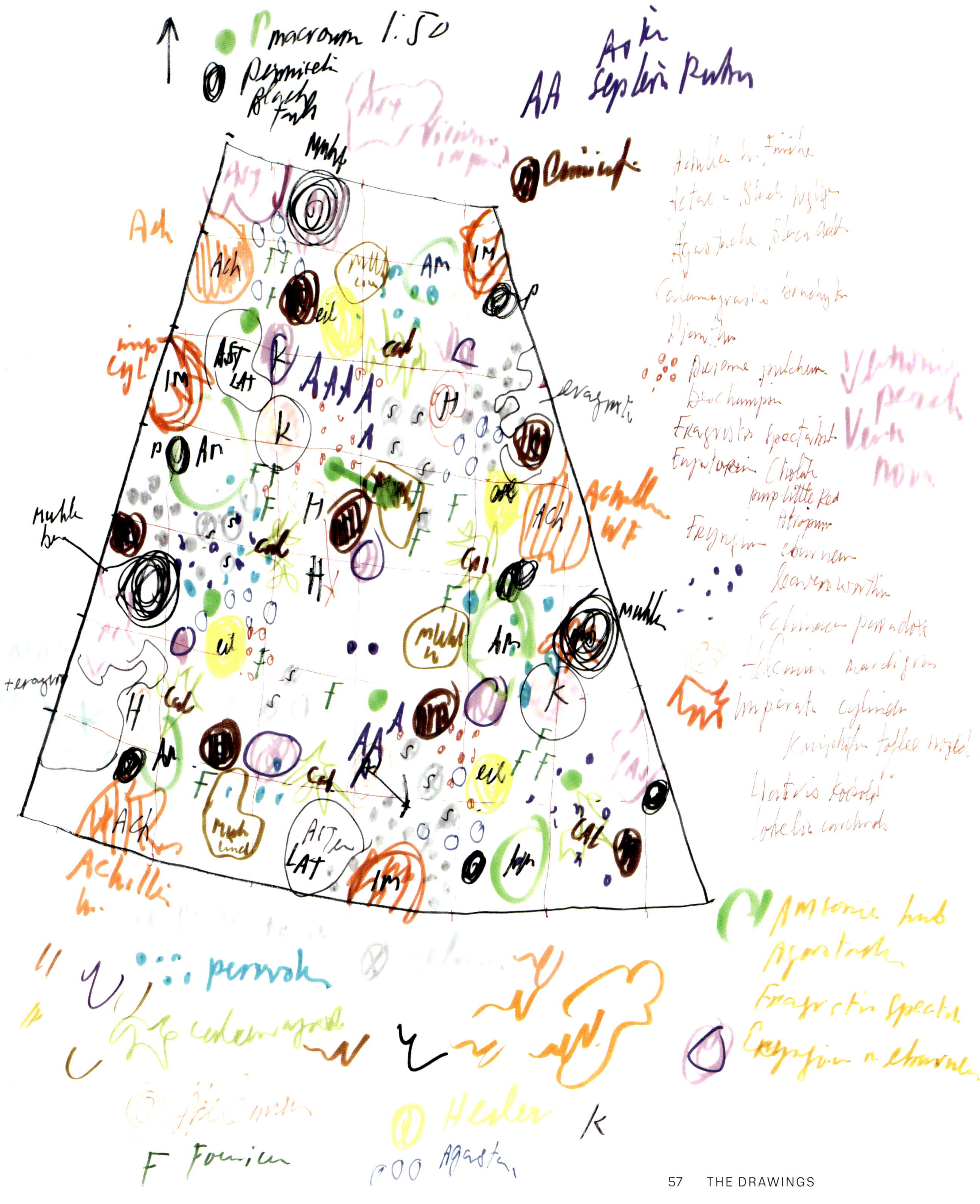

Preliminary sketches for the two terraces of an exhibition garden at the Palais de Tokyo, Paris, 2012. Sketches of this kind are about exploring basic distribution and outline relationships of key plant groups. The upper drawing shows the individual plants and small plant groups, with open spaces for larger plant groups. In the lower drawing, individual plants are black to assess the flow of the overall design, and their relationship and balance with the larger groups.

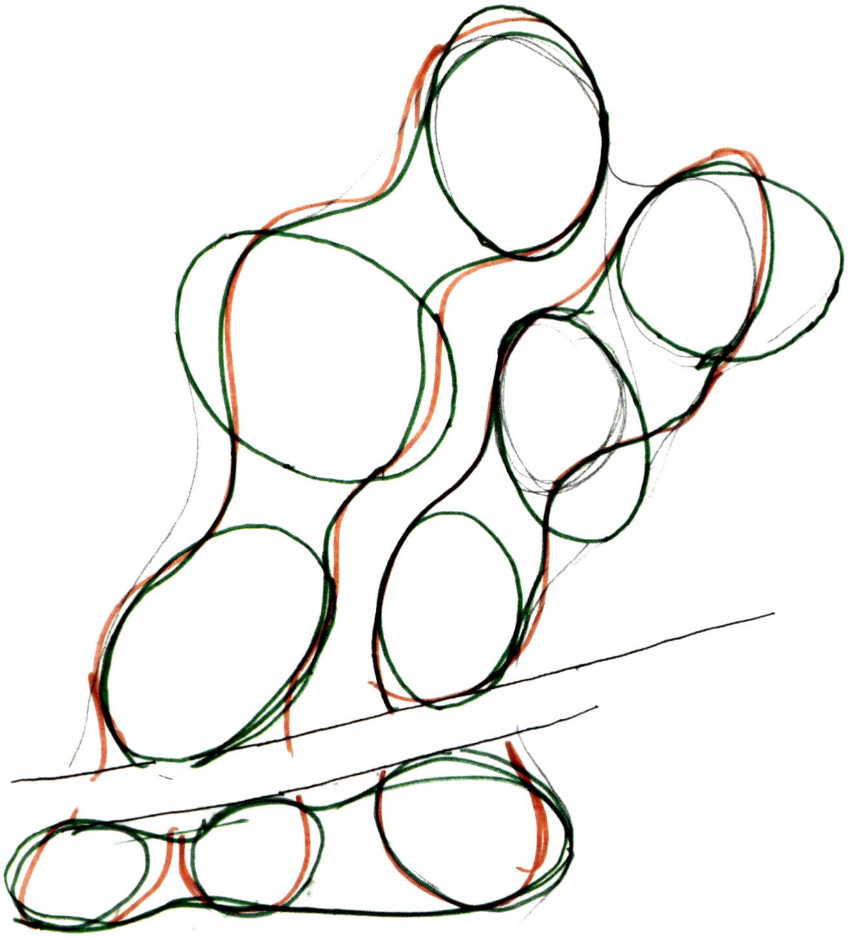

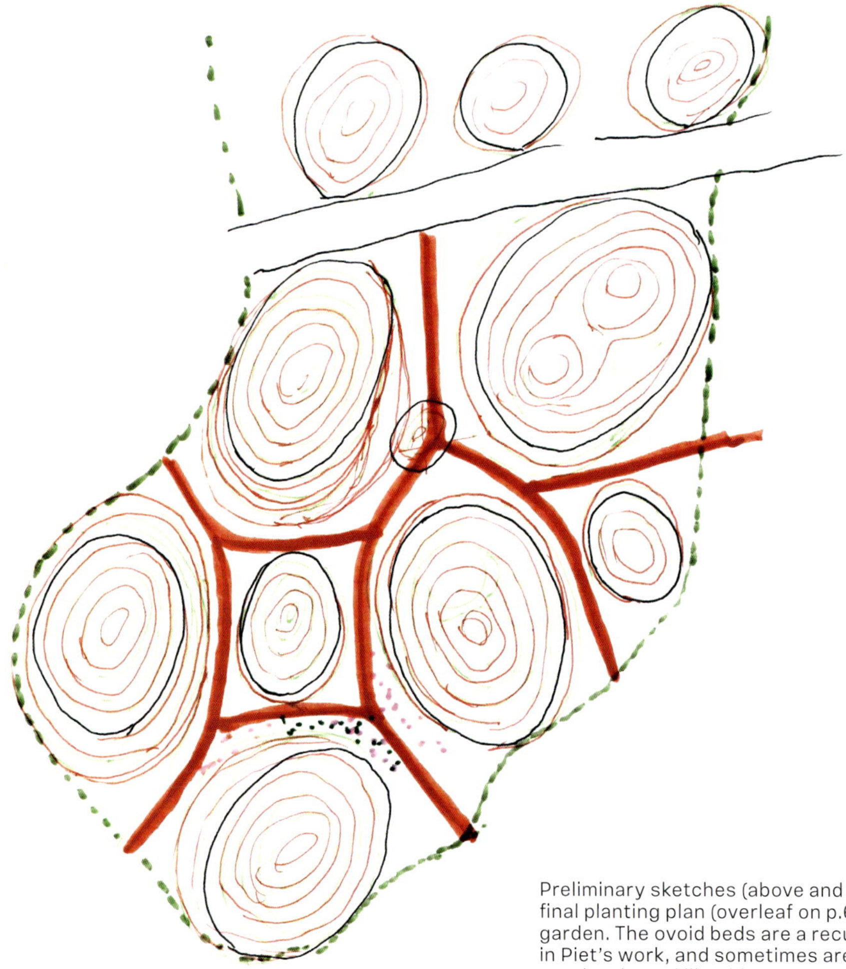

Preliminary sketches (above and right) and a final planting plan (overleaf on p.62) for a private garden. The ovoid beds are a recurring theme in Piet's work, and sometimes areas of grass may be shaped like this, as at Hauser & Wirth Somerset (see pp.132–35). The planting uses a *Sporobolus* matrix, with six other grass species in smaller quantities.

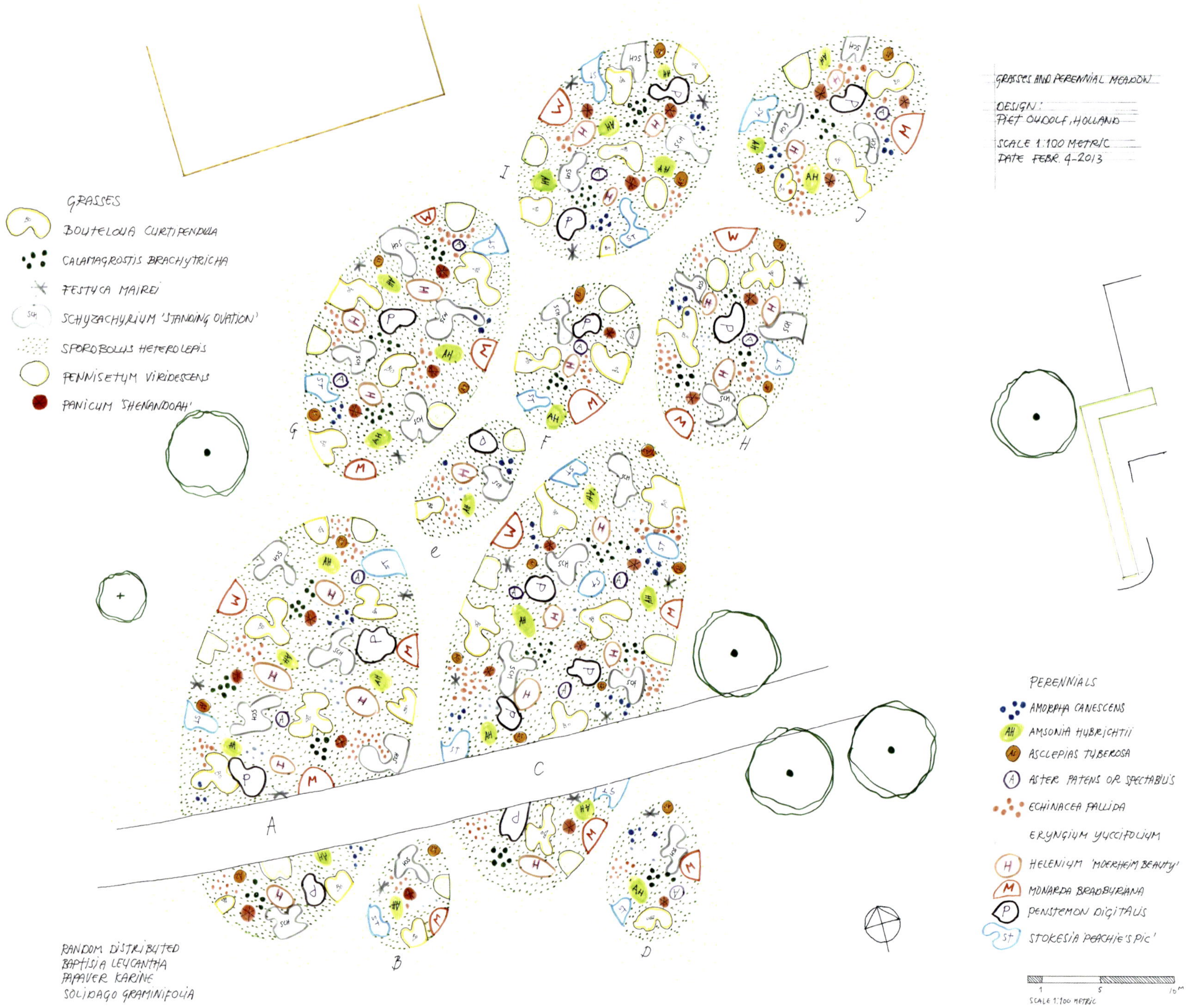

Matrix planting plan for a private garden (see previous pages), where oval islands use very versatile *Sporobolus heterolepis* as the matrix. It is worth noting the number of other grass species used here, whose presence will tend to blur and reduce the visual dominance of the *Sporobolus*.

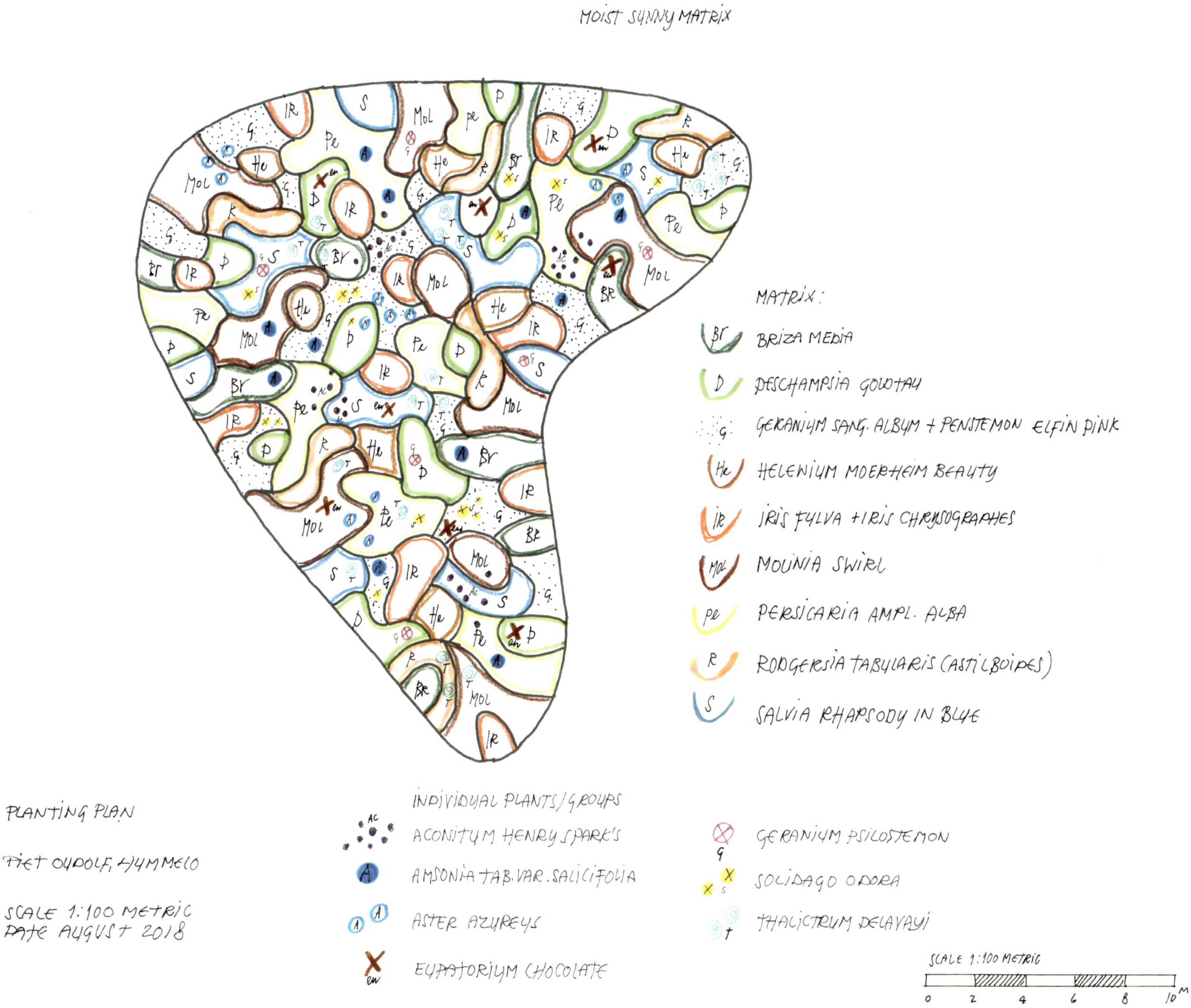

Plan for a block planting for a moist site in a private garden, 2018. Here perennials are arranged in blocks of one variety: however, a number of other varieties are also scattered throughout as individuals.

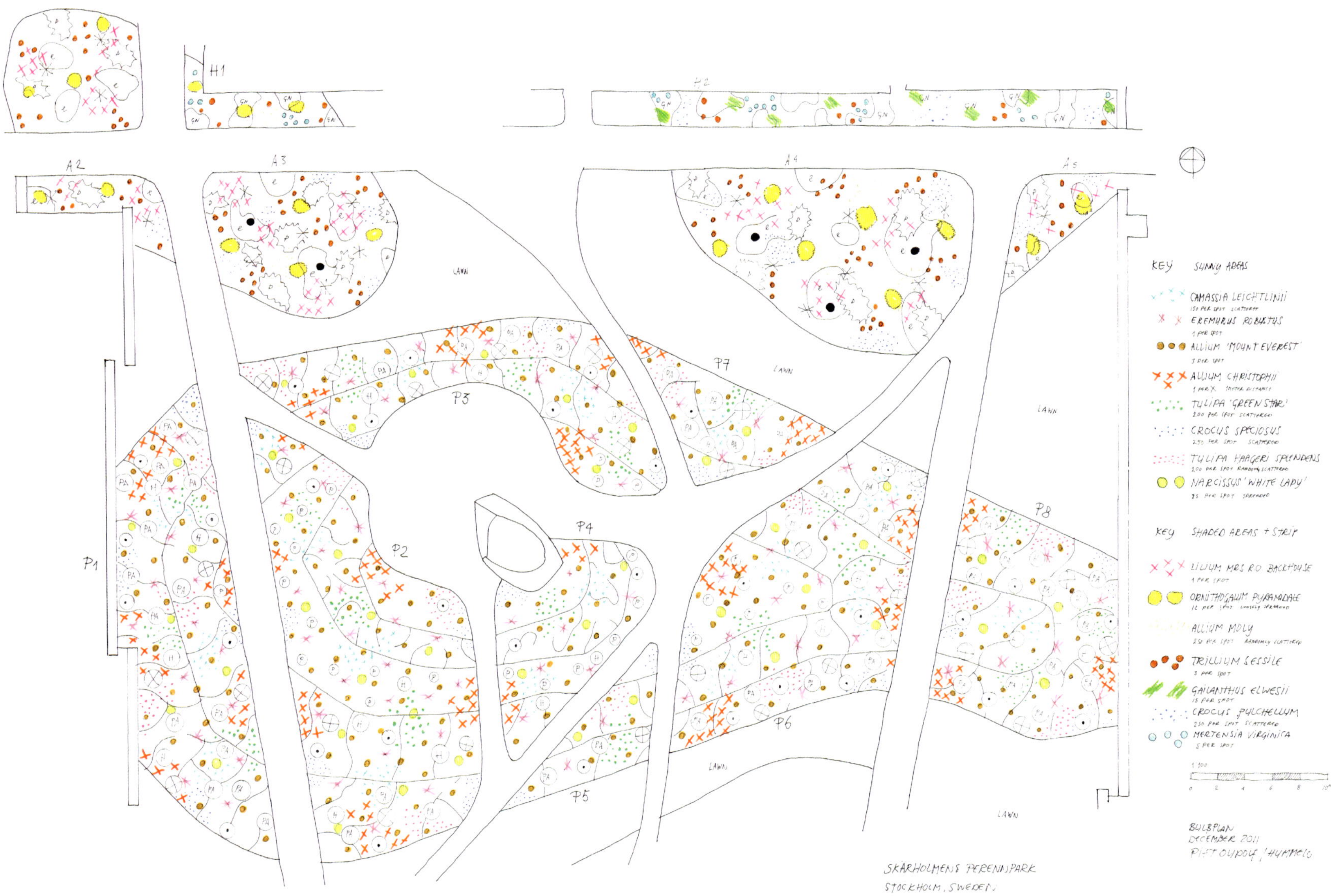

Bulb plan for Skärholmen public park, Stockholm, 2011. The perennial groups are organized in roughly concentric rings, but the pattern of bulb planting is independent of this.

The different phenologies of perennials and bulbs allow for two separate and independent plans.

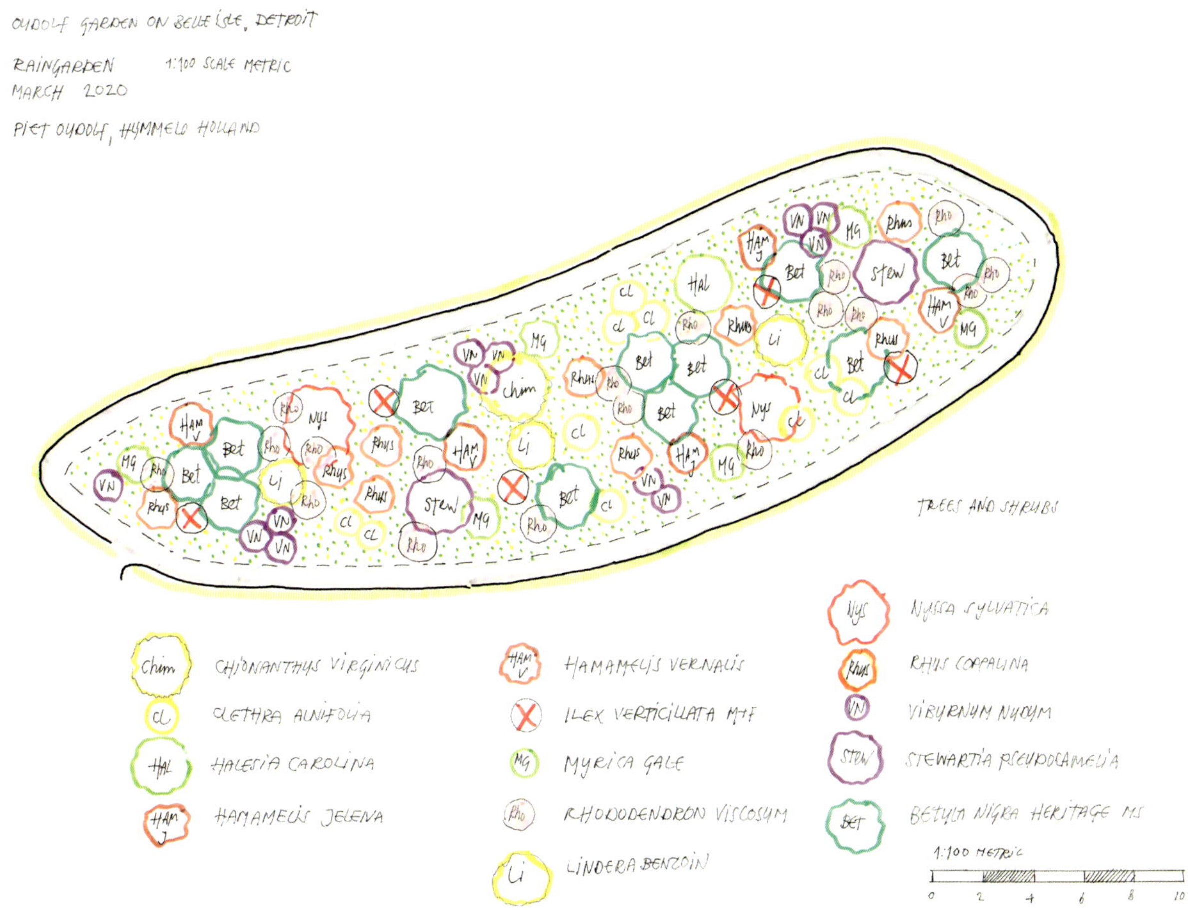

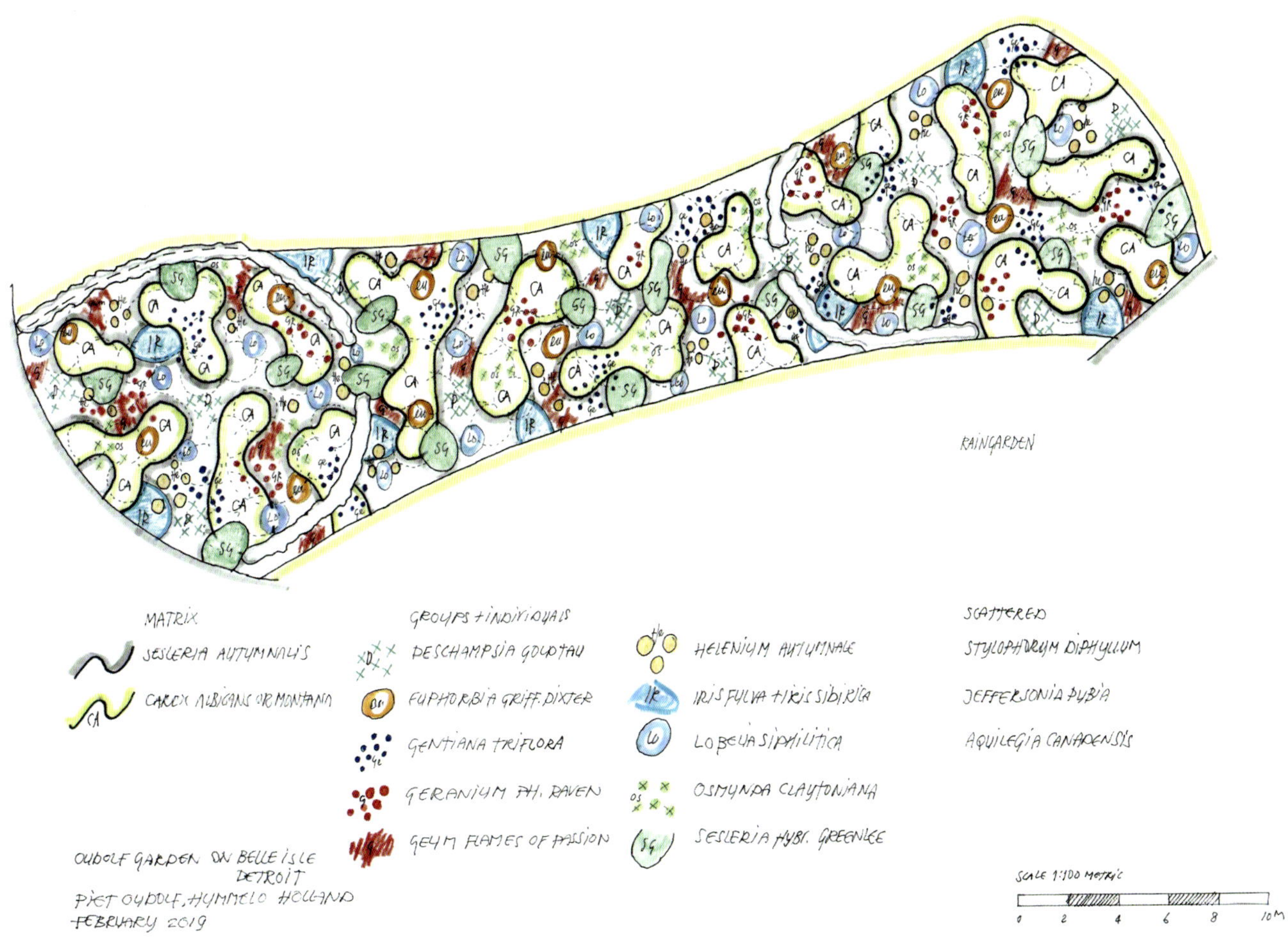

The Oudolf Garden, Detroit (see pp.226–27) is vulnerable to flooding in places; one lower area, next to the main perennial borders which are raised to protect against flooding, has been designed as a rain garden, with plants that can survive temporary inundation before the water soaks away: trees and shrubs plan (above), perennials plan (below), 2019–20.

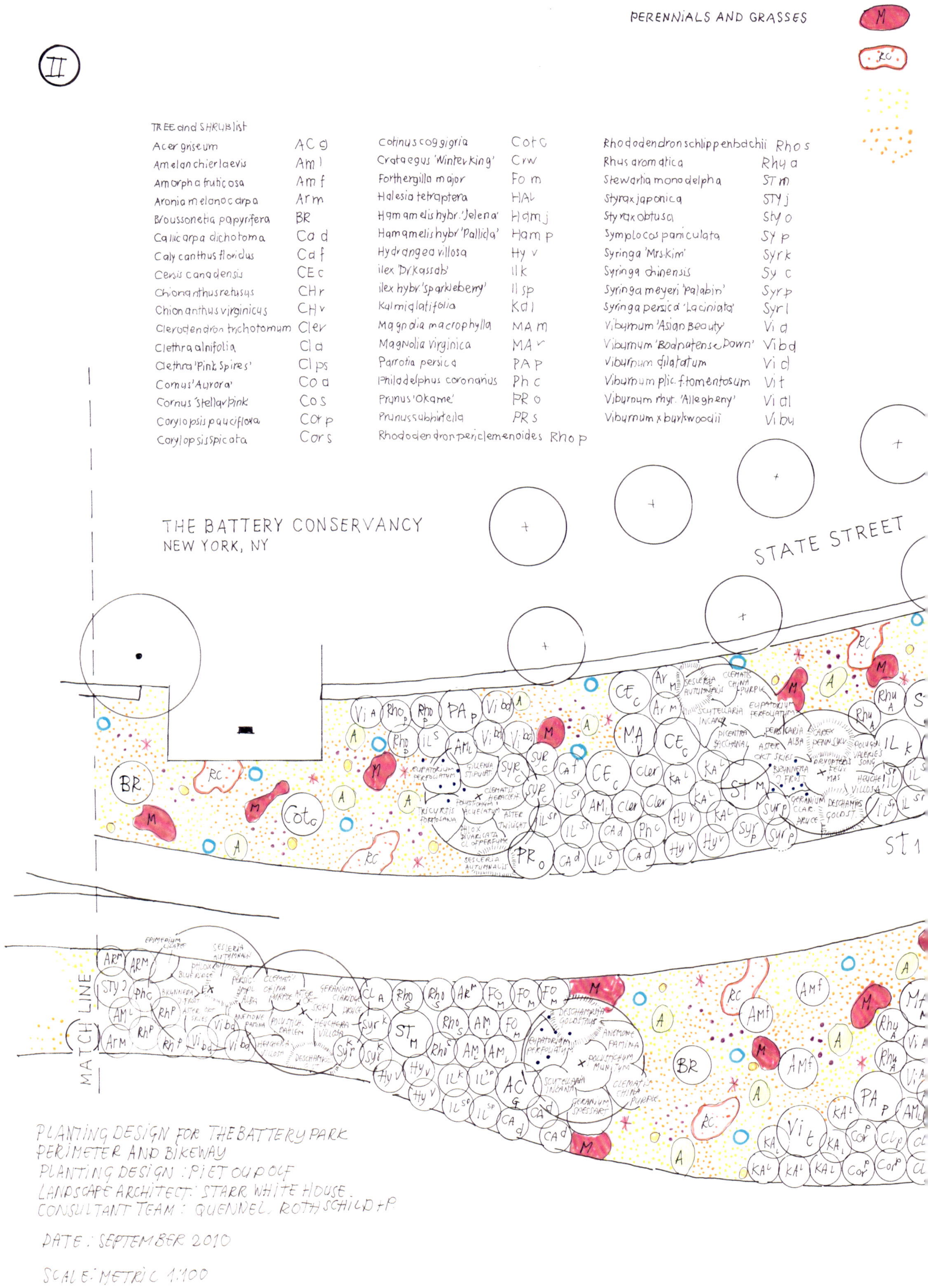
PERENNIALS AND GRASSES
M
RC
II
TREE and SHRUB list
Acer griseum ACg
Amelanchier laevis Aml
Amorpha fruticosa Amf
Aronia melanocarpa Arm
Broussonetia papyrifera BR
Callicarpa dichotoma Cad
Calycanthus floridus Caf
Cercis canadensis CEc
Chionanthus retusus CHr
Chionanthus virginicus CHv
Clerodendron trichotomum Cler
Clethra alnifolia Cla
Clethra 'Pink Spires' Clps
Cornus 'Aurora' Coa
Cornus 'Stellar Pink' Cos
Corylopsis pauciflora Corp
Corylopsis spicata Cors
Cotinus coggigria Cotc
Crataegus 'Winter King' Crw
Forthergilla major Fom
Halesia tetraptera HAL
Hamamelis hybr. 'Jelena' Hamj
Hamamelis hybr 'Pallida' Hamp
Hydrangea villosa Hyv
ilex 'Dr Kassab' ilk
ilex hybr 'Sparkleberry' ilsp
Kalmia latifolia Kal
Magnolia macrophylla MAm
Magnolia virginica MAv
Parrotia persica PAP
Philadelphus coronarius Phc
Prunus 'Okame' PRo
Prunus subhirtella PRs
Rhododendron periclemenoides Rhop
Rhododendron schlippenbachii Rhos
Rhus aromatica Rhya
Stewartia monodelpha STm
Styrax japonica STYj
Styrax obtusa Styo
Symplocos paniculata Syp
Syringa 'Mrs Kim' Syrk
Syringa chinensis Syc
Syringa meyeri 'Palabin' Syrp
Syringa persica 'Laciniata' Syrl
Viburnum 'Asian Beauty' Via
Viburnum 'Bodnatense Dawn' Vibd
Viburnum dilatatum Vid
Viburnum plic. f. tomentosum Vit
Viburnum rhyt. 'Allegheny' Vial
Viburnum x burkwoodii Vibu
THE BATTERY CONSERVANCY
NEW YORK, NY
STATE STREET
MATCH LINE
PLANTING DESIGN FOR THE BATTERY PARK
PERIMETER AND BIKEWAY
PLANTING DESIGN : PIET OUDOLF
LANDSCAPE ARCHITECT: STARR WHITEHOUSE
CONSULTANT TEAM : QUENNEL ROTHSCHILD+P
DATE : SEPTEMBER 2010
SCALE: METRIC 1:100

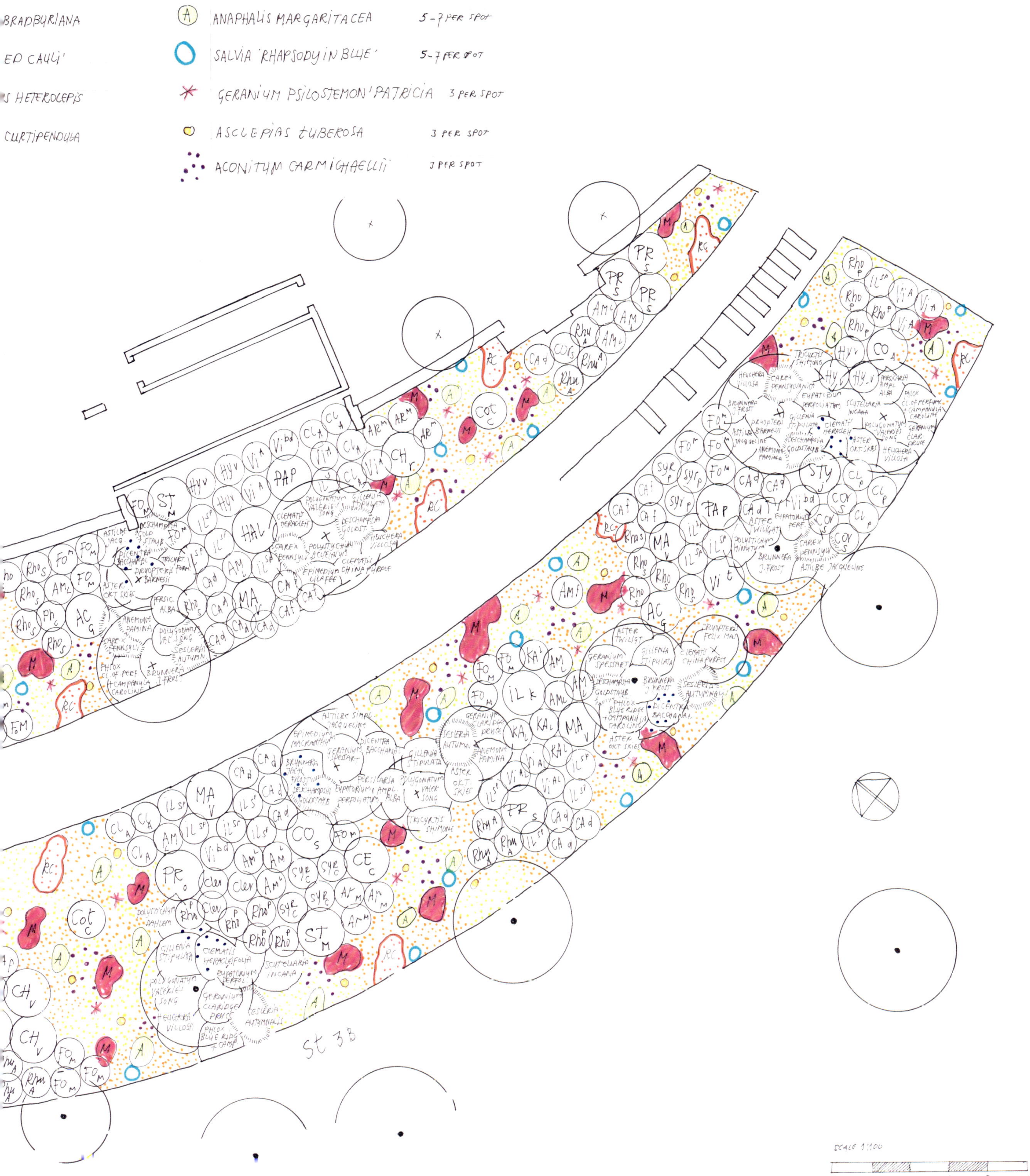

Tree- and shrub-dominated planting for the perimeter and bikeway on the street side of Battery Park, New York, 2010. Trees play an important role here, underplanted with perennial groups. More open areas have some matrix plantings, with two native grass species forming the matrix.

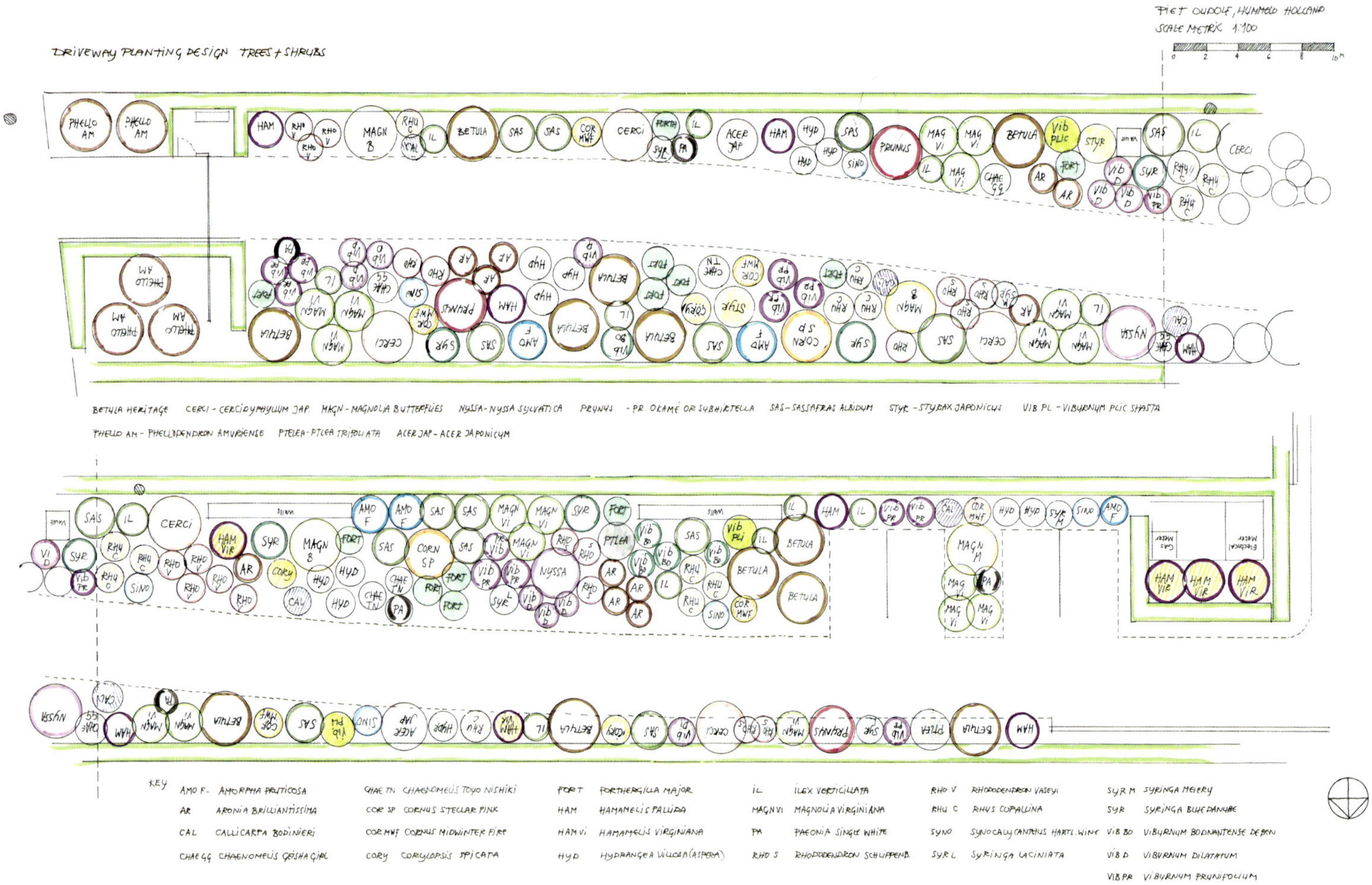

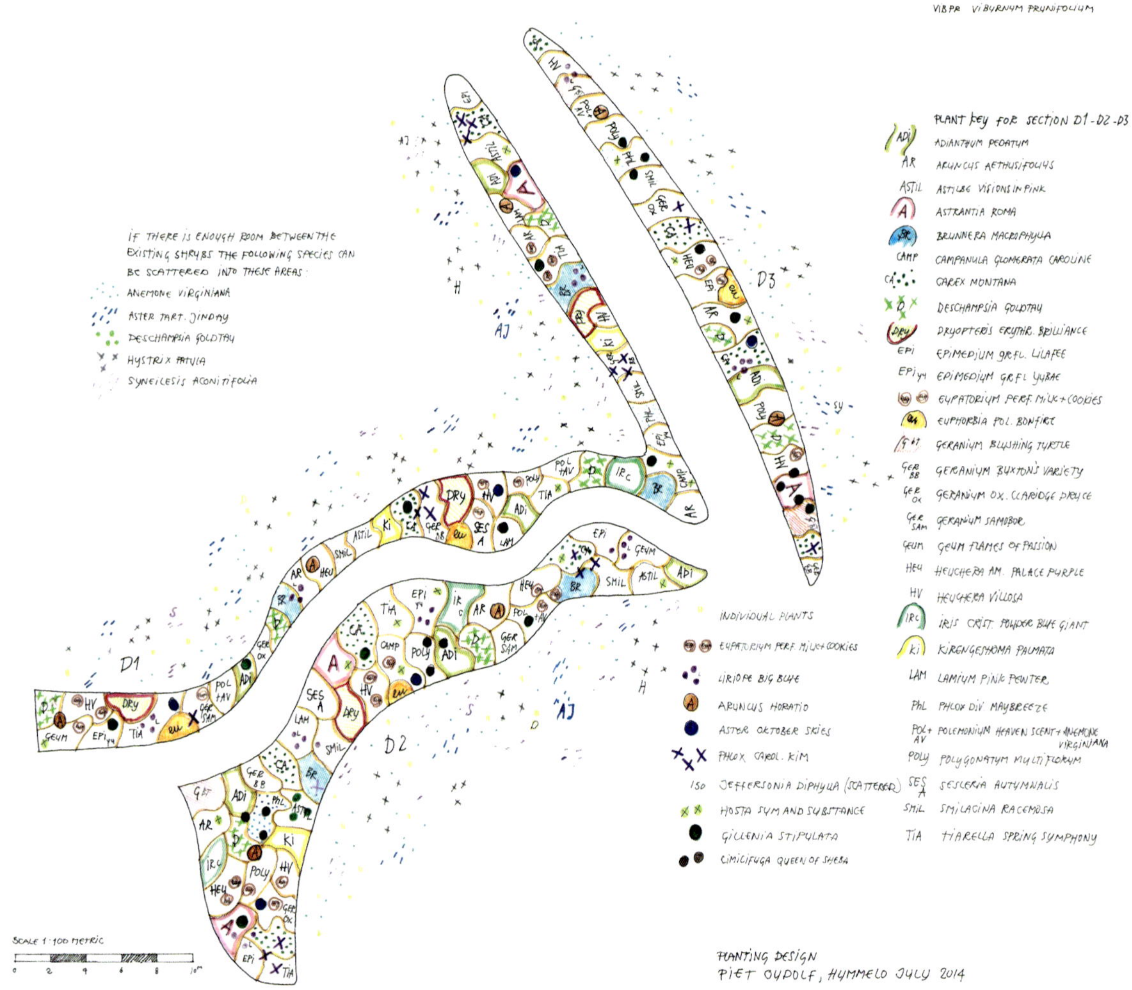

Two examples of tree and shrub planting for private gardens. The upper plans show shrubs and small trees for lining a driveway: species such as *Cornus*, *Hydrangea* and *Viburnum*. In the lower plan, woodland species are planted beneath existing shrubs.

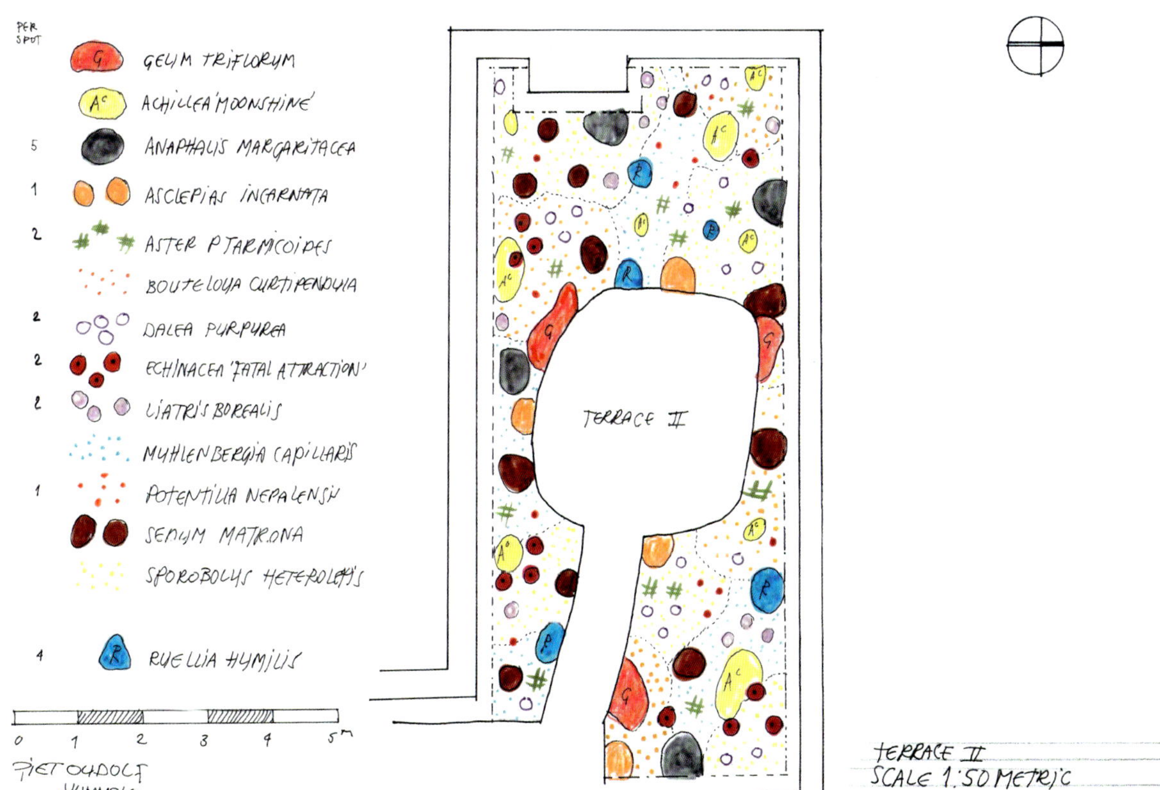

Preliminary sketch and final plan for a small private roof terrace, 2011. This design uses three species of grass forming separate areas of matrix planting, each dominated by one species.

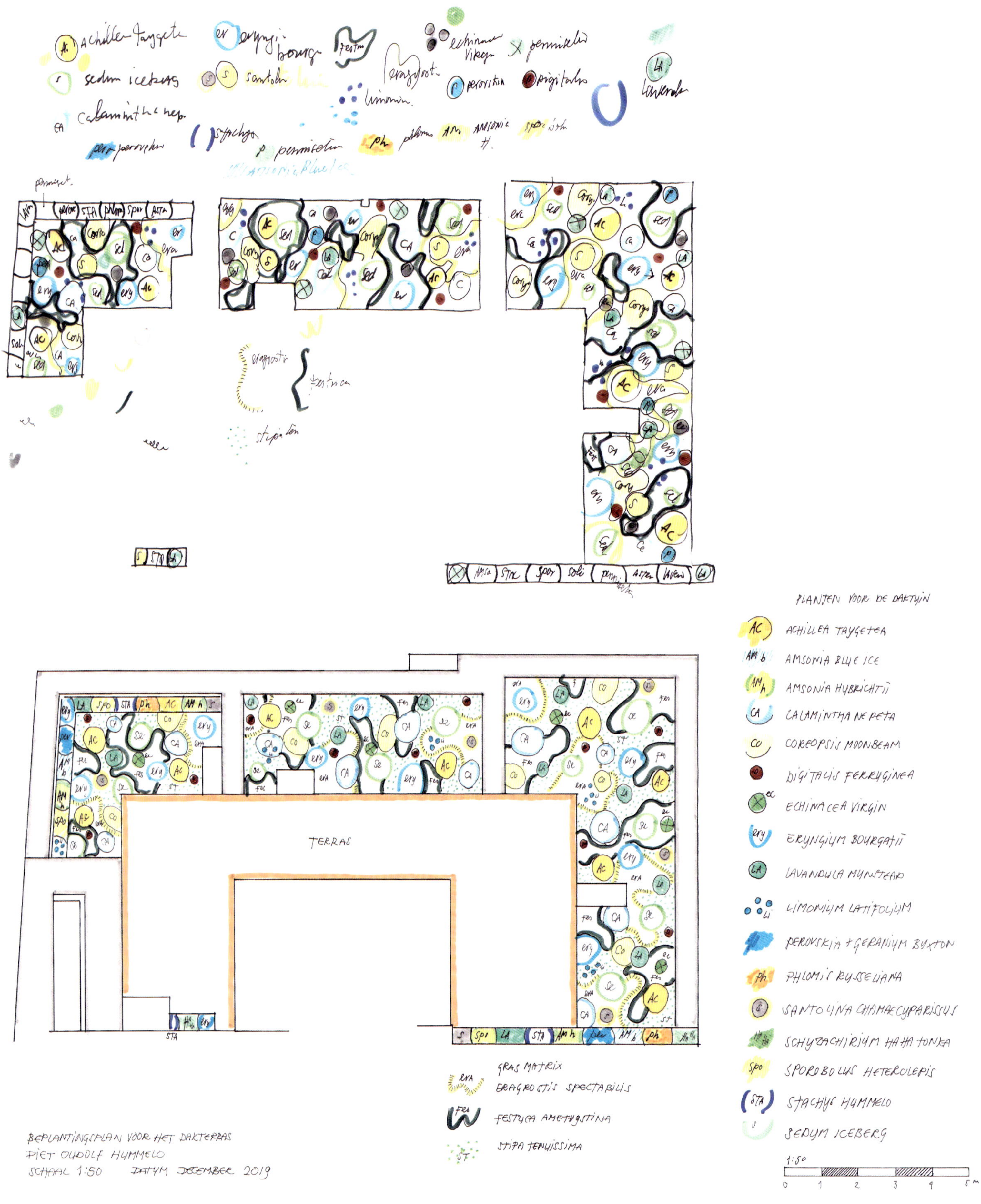

Preliminary sketch and final plan for a roof terrace, 2019. While not a matrix planting, its use of *Stipa tenuissima* will create the impression of a grassy meadow. The substrate here is 15–25 centimetres (6–10 in) deep, normal for green roofs but very stressful for the plants. This grass species has a very limited water requirement.

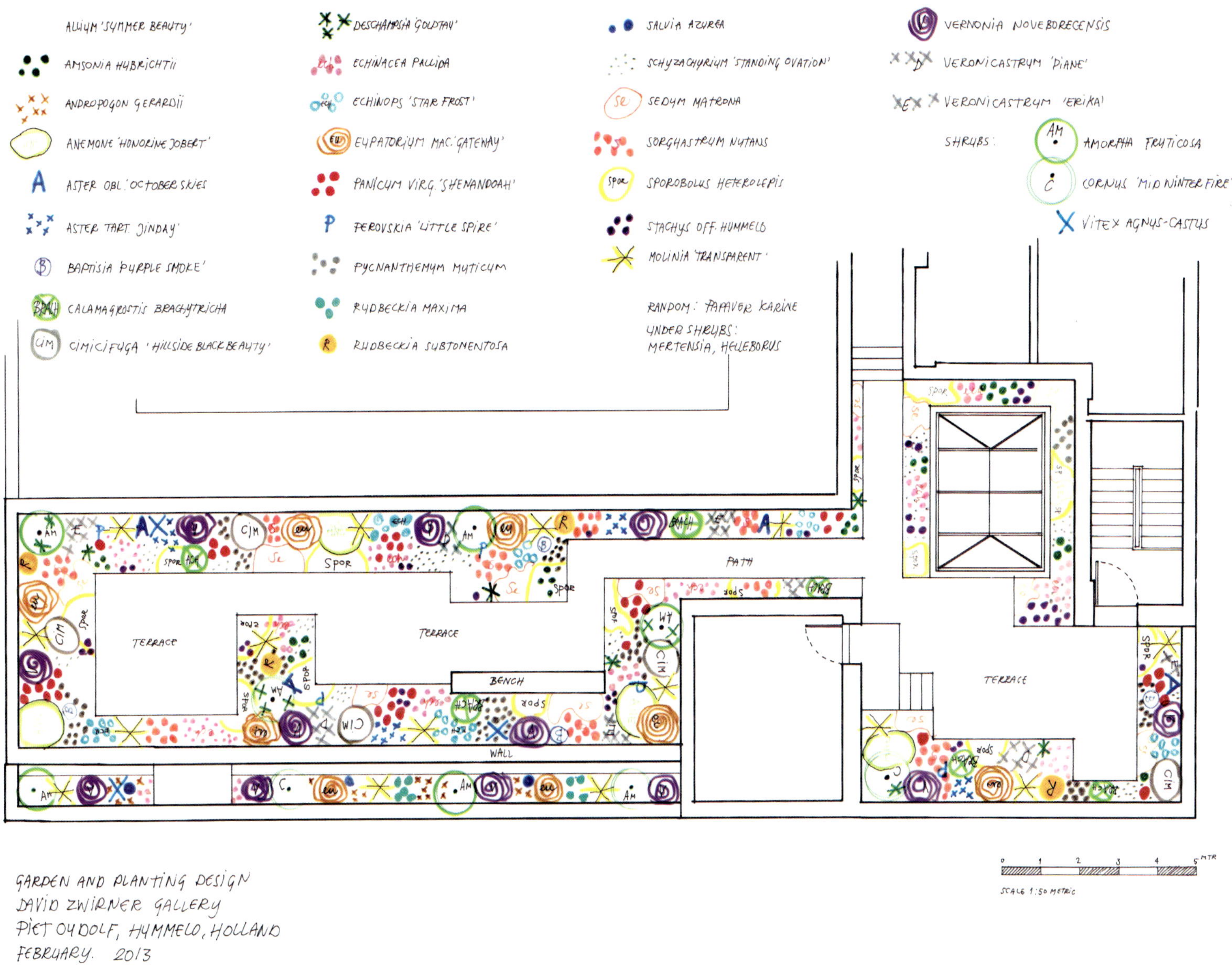

Planting plan for a roof terrace at David Zwirner contemporary art gallery in New York, 2013. Tight spaces create certain design challenges: all the plants are highly visible, and so all have to perform.

A garden
isn’t a landscape painting
that you look at,
but a dynamic process
that’s always changing.

Piet Oudolf

THE GARDENS

Lurie Garden
Millennium Park, Chicago, Illinois, USA
2001–04

The Lurie Garden opened in July 2004 and is a garden area within a larger, public space – Millennium Park – created in collaboration with landscape architects Gustafson Guthrie Nichol with the support of the Ann and Robert H. Lurie Foundation. This collaboration involved a conceptual framework being developed for the garden, not normally an aspect of Piet's work, and one that arguably adds nothing to the public perception of what has become an immensely popular part of the Chicago waterfront landscape, and which has also undoubtedly helped stimulate growing interest in garden-making in the region. Surrounded by a high 'shoulder' hedge, a broad boardwalk (symbolizing the wooden sidewalks that once lined the city's streets) and a narrow waterway ('the seam') divide the site into two distinct regions of planting: the 'dark plate' and the 'light plate'. The dark plate is an area of open woodland, richly underplanted with shade-tolerant plants symbolizing the wild landscape that existed before the arrival of white settlers. Considerably larger, and set at a lower level, is the more open area of the light plate (see left and pp.80–81).

The entire site is, in fact, a roof planting over a car park; the depth of soil varies, so that there are constant slight changes in elevation. A dominant feature in early summer is the Salvia River, where four cultivars of European meadow *Salvia* are used to create a dramatic drift of purple-blue; Piet does not like repeating himself, but this is a copy of a feature he made for his first public park, in Enköping, Sweden (see p.247). A bulb planting by Dutch designer Jacqueline van der Kloet was added a year after installation.

Around half the species used are Midwest natives. Block planting is used throughout, apart from one small area, 'the meadow' (see p.80) – an area of intermingled planting that is more consciously naturalistic, with a higher concentration of grasses, and more intermingling of varieties. Although not described as matrix planting at the time, it was Piet's first foray into a freer and less controlled spatial arrangement of plants.

Soft white *Parthenium integrifolium* and pink *Echinacea pallida* are two of the locally native plants that adorn the gently rolling terrain of Chicago's Lurie Garden. For many Chicagoans they are the first sight they may have of these prairie species. *Overleaf*: The original planting plan for the Lurie Garden, 2001, with Piet's handwritten annotations.

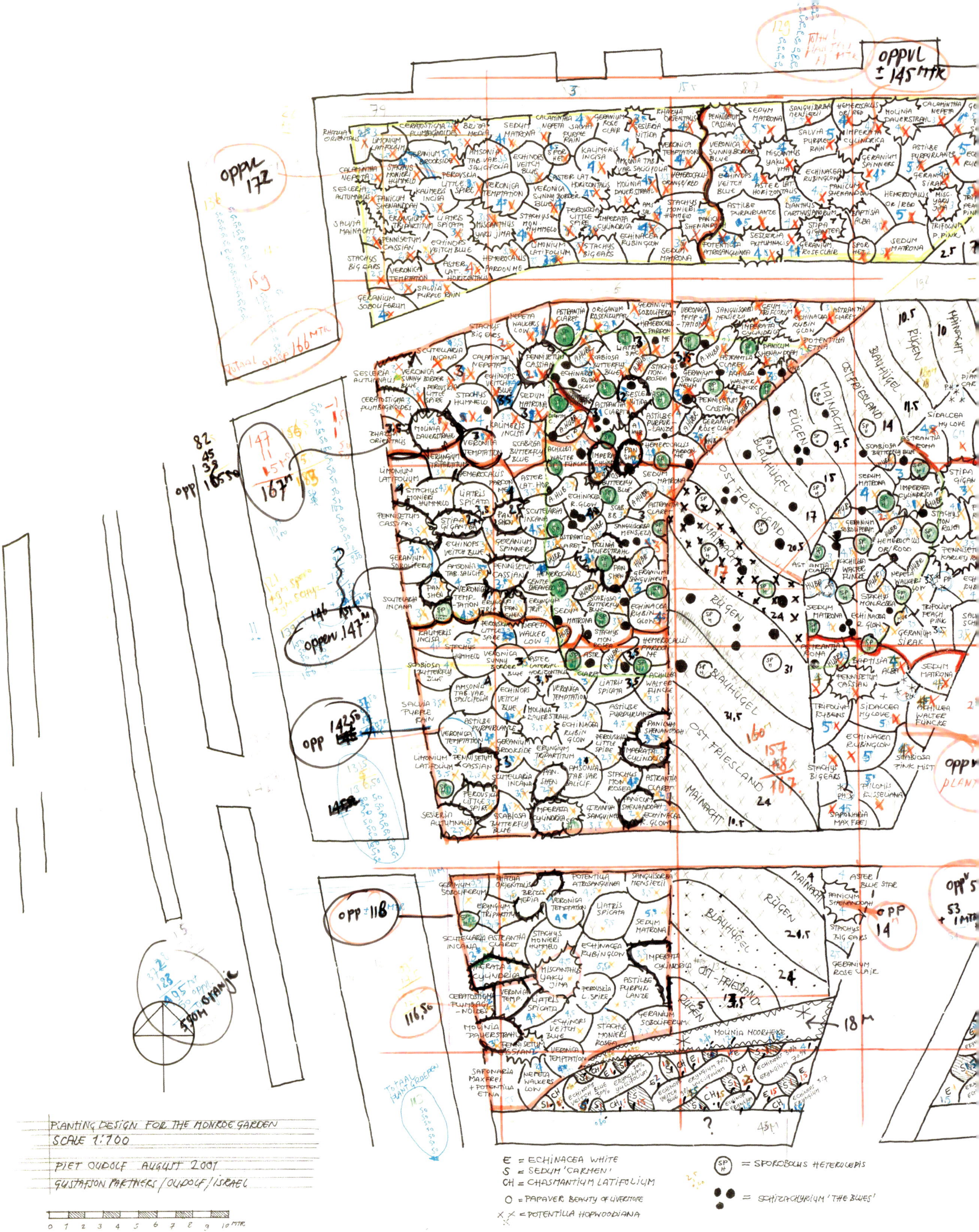
PLANTING DESIGN FOR THE MONROE GARDEN
SCALE 1:700
PIET OUDOLF AUGUST 2001
GUSTAFSON PARTNERS / OUDOLF / ISRAEL
0 1 2 3 4 5 6 7 8 9 10 MTR
E = ECHINACEA WHITE
S = SEDUM 'CARMEN'
CH = CHASMANTIUM LATIFOLIUM
O = PAPAVER BEAUTY OF LIVERMORE
X X = POTENTILLA HOPWOODIANA
SP H = SPOROBOLUS HETEROLEPIS
= SCHIZACHYRIUM 'THE BLUES'

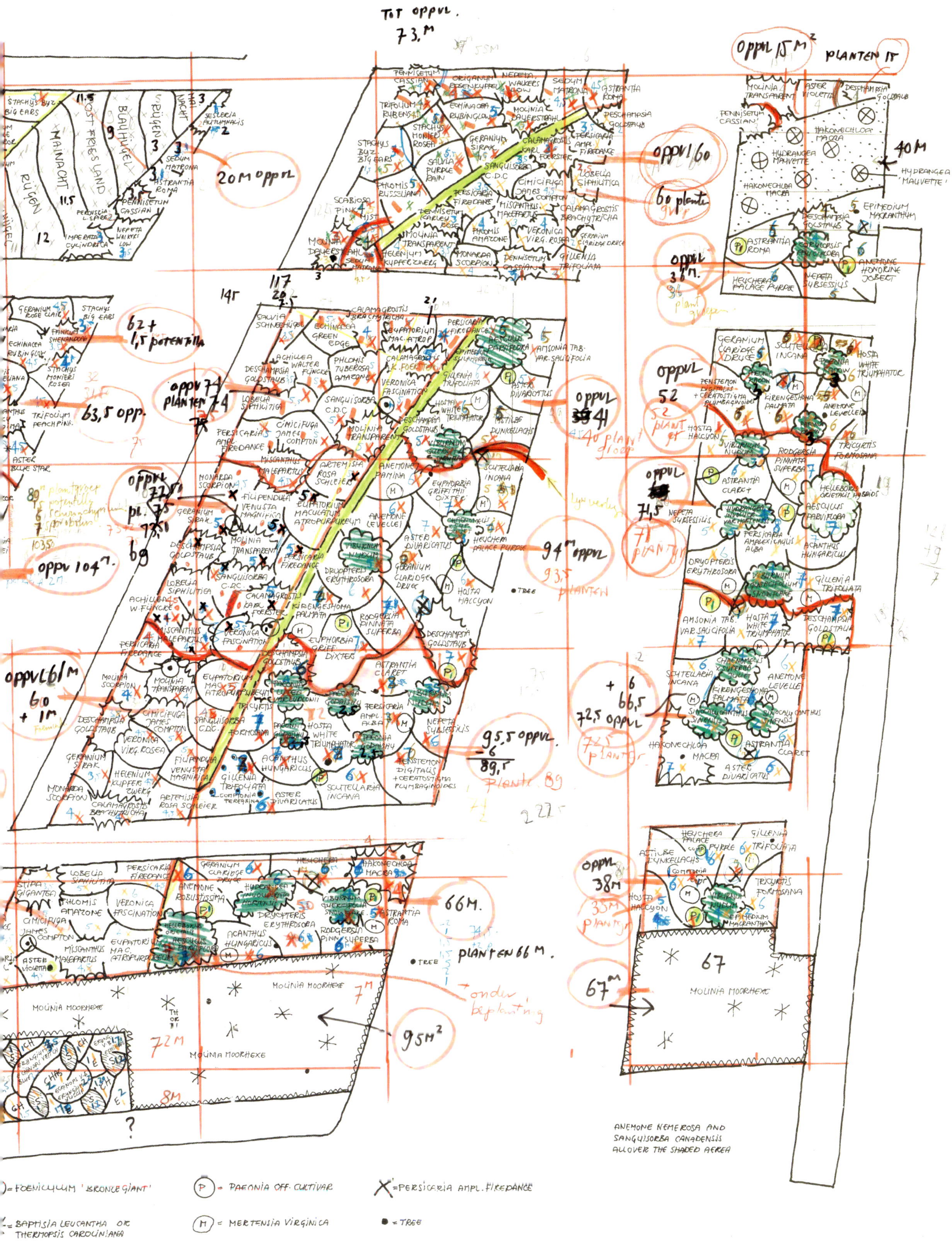
TOT OPPVL 73 M
OPPVL 15 M²
PLANTEN 15
40 M
20 M OPPVL
OPPVL 60
60 plants
OPPVL 104
OPPVL 61 M
66 M.
PLANTEN 66 M.
95 M²
72 M
67 M
MOLINIA MOORHEXE
ANEMONE NEMEROSA AND SANGUISORBA CANADENSIS ALLOVER THE SHADED AEREA
= FOENICULUM 'BRONZE GIANT'
P = PAEONIA OFF. CULTIVAR
= PERSICARIA AMPL. FIREDANCE
= BAPTISIA LEUCANTHA OR THERMOPSIS CAROLINIANA
M = MERTENSIA VIRGINICA
• = TREE

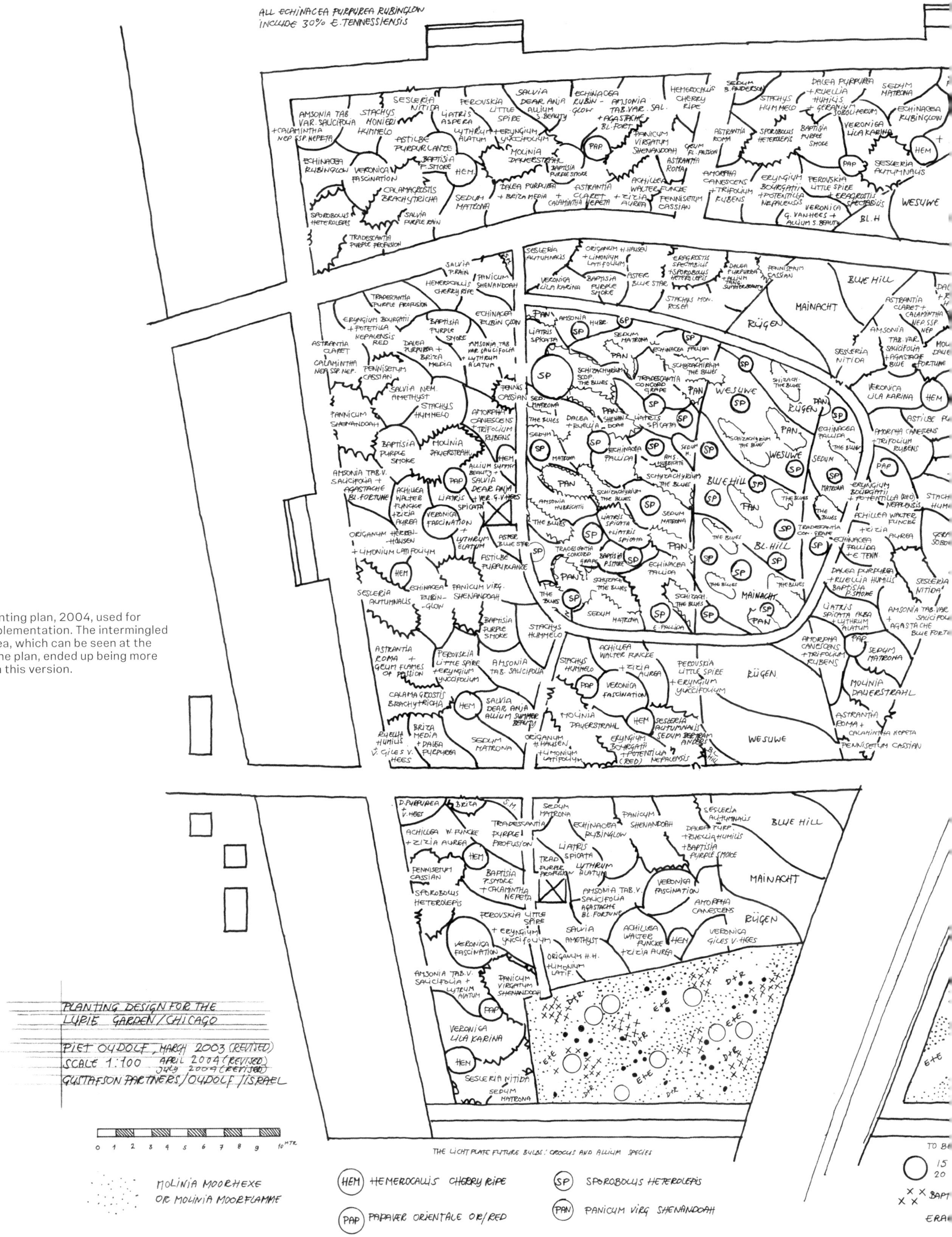

Revised planting plan, 2004, used for the final implementation. The intermingled meadow area, which can be seen at the bottom of the plan, ended up being more extensive in this version.

Pi PAEONIA J.V. LEEUWEN
D DIGITALIS FERRUGINEA
E+E = ECHINACEA WHITE + ERYNGIUM YUCCIFOLIUM.
D+R = DALEA PURP. ALBA + RUELLIA HUMILIS (MOVING THROUGH THE PLANTING)
SILPHIUM LACINIATUM
THE DARK PLATE: IN THE SECOND YEAR TO BE UNDERPLANTED WITH: ANEMONE NEMOROSA SANGUINARIA CANADENSIS, MERTENSIA AND TRILLIUM.
MOLINIA DAUERSTRAHL
ASTRANTIA ROMA + GEUM FL. OF PASSION
WESUWE
ORIGANUM HERRENHAUSEN + LIMONIUM LATIFOLIUM
BLUE HILL
MAINACHT
GERANIUM SOBOLIFERUM
PANICUM SHENANDOAH + ERYNGIUM YUCCIFOLIUM
ASTER BLUE STAR
STACHYS MON. ROSEA
SALVIA NEM. AMETHYST
TRADESCANTIA PURPLE PROFUSION
LIATRIS
ACHILLEA W. FUNCKE + ZIZIA AUREA
SALVIA AMETHYST
PEROVSKIA LITTLE SPIRE + ERYNGIUM YUCCIFOLIUM
ERAGROSTIS SPECT.
SEDUM MATRONA
HEMEROCALLIS G. SHEPHERD
VERONICA EVELYNE
SESLERIA NITIDA
HELENIUM RUBINZWERG
MISCANTHUS FERNER OSTEN
EUPATORIUM PURPLE BUSH
SANGUISORBA RED THUNDER
PHLOMIS TUB. AMAZONE
MONARDA SCORPION
ASTER VIOLETTA
PHLOX PANICULATA
SAPONARIA MAX FREI + SEDUM MATRONA
ECHINOPS SPHAEROCEPHALUM
SORGHASTRUM SIOUX BLUE
ASTILBE PURPURLANZE
HAKONECHLOA MACRA
MOLINIA LIT. TRANSPARENT
ASTER OKT. SKIES + EUPATORIUM GATEWAY
CIMICIFUGA J. COMPTON
SANGUISORBA R. THUNDER + S. MENZIESII
CAMPANULA GLOM. CAROLINE + ASTR. ROMA
TRICYRTIS FORMOSANA
HELLEBORUS ORIENTALIS PINK
HOSTA BLUE ANGEL
EPIMEDIUM LILAFEE
ANEMONE LEVELLEI
VIBURNUM N404M
CERATOSTIGMA PLUMBAGINOIDES
SCUTELLARIA INCANA
KIRENGESHOMA PALMATA
DESCHAMPSIA GOLDSTAUB
GILLENIA TRIFOLIATA
ASTRANTIA ROMA + GEUM FLAMES OF PASSION
ASTER DIVARICATUS
PERSICARIA AMPL. FIRETAIL + GERANIUM PHAEUM ALBUM
RODGERSIA PINNATA SUPERBA
SMILACENA RACEMOSA
AESCULUS PARVIFLORA
EPIMEDIUM GR. FLORUM
ACANTHUS HUNGARICUS + ARUNCUS HORATIO
DATISCA CANNABINA
HOSTA WHITE TRIUMPHATOR
PAEONIA LUTEA
CAREX MUSKINGUMENSIS
HELLEBORUS ORIENTALIS WHITE
RODGERSIA PINNATA SUPERBA (HENRICI)
EPIMEDIUM GRANDIFLORUM
POLYSTICHUM HERRENHAUSEN
HOSTA HALCYON
TRICYRTIS FORMOSANA
HEUCHERA PALACE PURPLE
HYDRANGEA INVOLUCR. HORTENSIS
ANEMONE HUPEHENSIS SPLENDENS
GILLENIA TRIFOLIATA

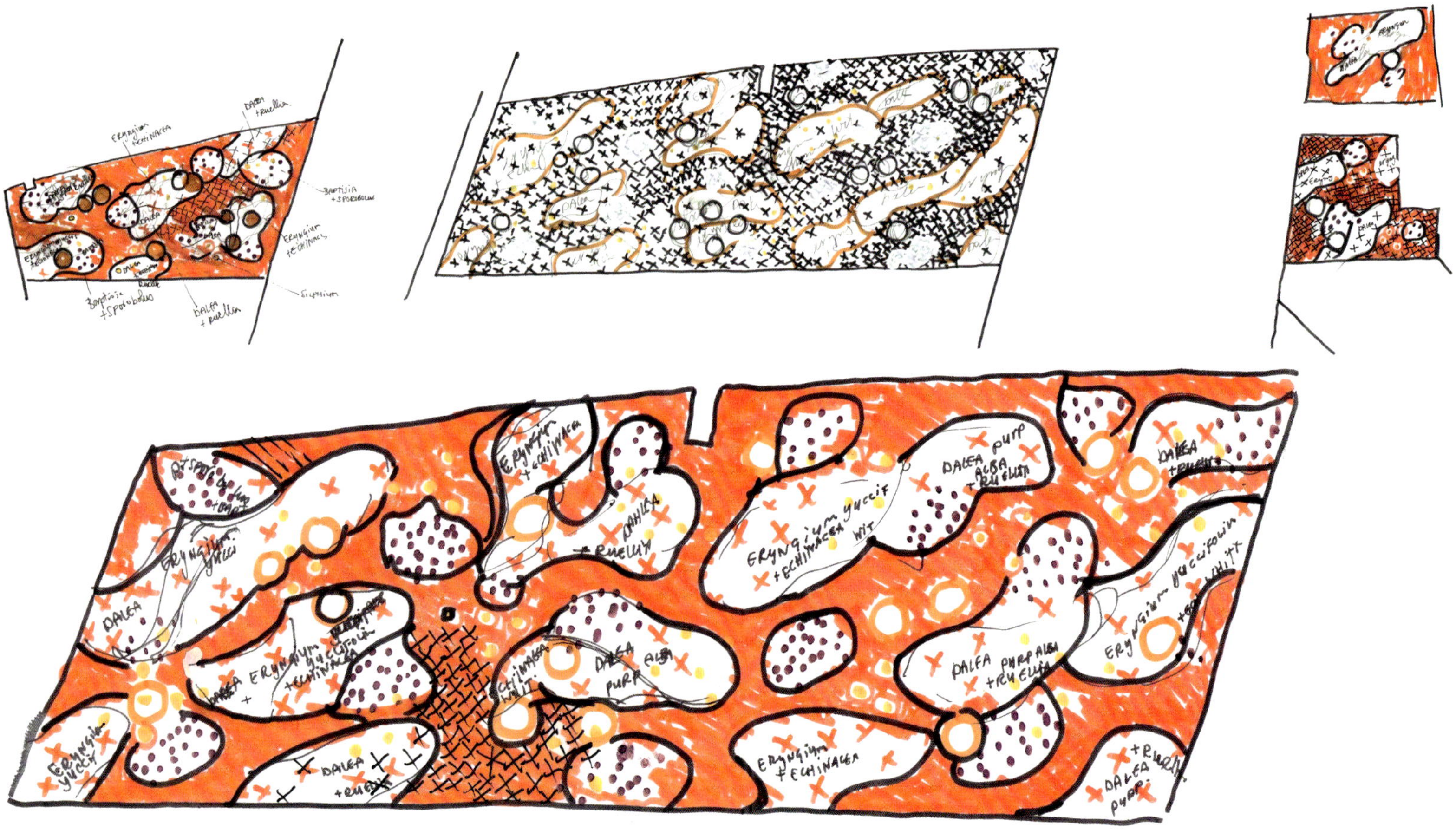

The Lurie Garden, clockwise from upper left: the Salvia River in early summer; *Echinacea pallida* in mid-summer; autumn colour with the yellow of *Amsonia hubrichtii* prominent; first sketches for the meadow area.

JOHN W. WILLIAMS, INC.
Weichsel Beef

The High Line

New York City, New York, USA
2006–

Piet's best-known project, the 2.3 kilometre (1.45 mile) former elevated freight railway on Manhattan's western side, was built in a series of phases that opened in 2009, 2011, 2014 and 2019, with a further phase opening in 2023. The High Line runs between buildings and sometimes through them, and much of the visitor experience is to do with the dramatic juxtapositions of the natural and the urban. The overall planting concept and approach was developed by Piet in collaboration with James Corner Field Operations, with Piet focusing on the specific design and detail of each planting area (see also p.96). In the first two phases, two distinct planting styles are used, largely determined by light: one where native grasses tend to dominate and the other in which shade features, partly determined by surrounding buildings but also extensive use of shrubs and small trees. These woodland areas have a perennial understorey that reflects the graphic, blocky nature of plant spread often found in forest-floor plants. The open, grassy areas, however, are among the most naturalistic of Piet's work anywhere, with high levels of variety intermingling, and matrix planting using locally native grass species (see for example, the Chelsea Grasslands as seen on pp.94–95). These planting styles present a stylized vision of two key natural habitats found in the wider northeastern region of the United States.

Parts of the High Line may look remarkably like a natural woodland-edge habitat, but in fact they receive a high level of maintenance from garden staff with support from a team of volunteers. Management is directed very much at maintaining a harmonious balance of species. The narrow width of the planting makes for a radically different visual environment – usually Piet's plantings rely on depth and scale to create impact. The result is a far more fine-textured and nuanced planting, and one that is, in fact, more naturalistic than most of his designs. A number of locations offer opportunities for pausing, so that visitors can appreciate different perspectives of the planting, for example the Spur at 30th Street (see pp.88–89), which uses a palette of trees, woodland shrubs and perennials. Here an open piazza with cascading wooden seating and panoramic views provides a space for public events and for visitors to interact with their surroundings. Two large 'tilted' planters create a dramatic planted threshold into the Spur and provide a lush wall of greenery after coming through the Coach Passage or from the streets of Midtown below.

The New York High Line cuts a swathe through the urban jungle; this is its start, at Gansevoort Street.

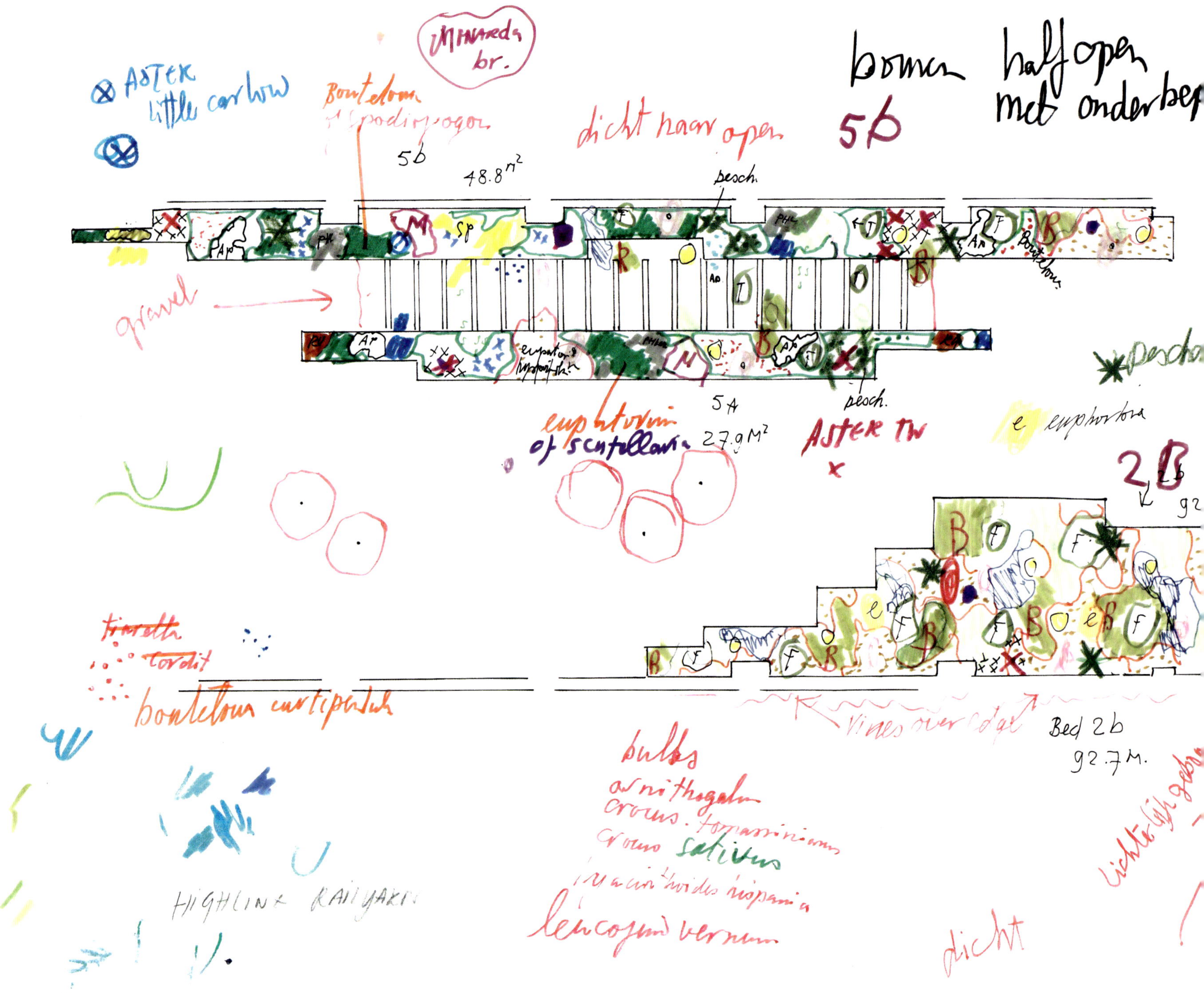

Preliminary sketches for the High Line, Rail Yards Section, phase 1. These early sketches were about envisioning the totality of the vegetation, but for planning and implementation purposes they had to be broken down into their component visual and functional layers: shrubs, block planting, scatter plants etc.

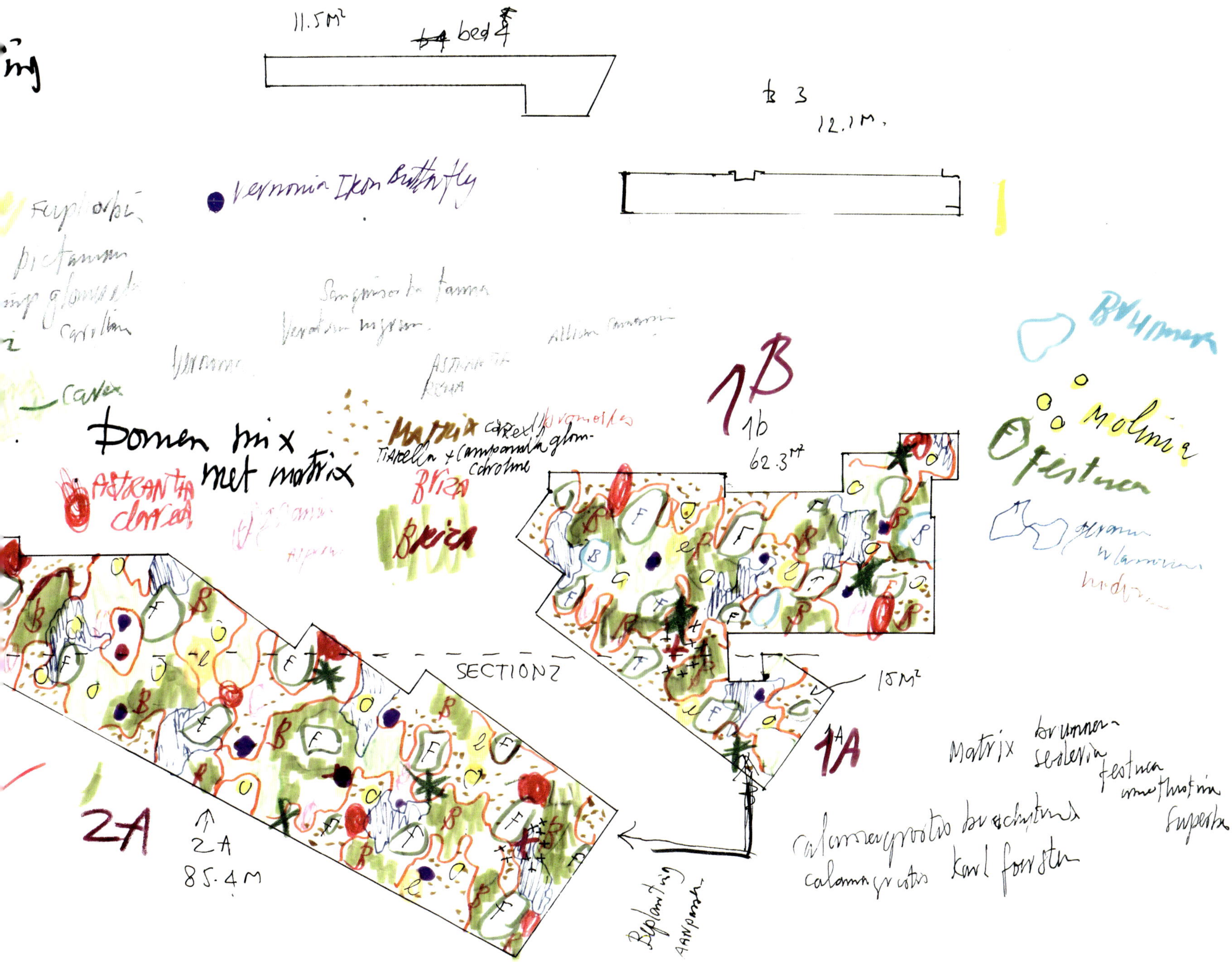
11.5M²
bed 4
12.1M
Vernonia Ikon Butterfly
1B
1b
62.3M²
Bomen mix
met matrix
Astrantia
Briza
SECTION 2
15M²
1A
2A
2A
85.4M
Matrix
Calamagrostis brachytricha
Calamagrostis Karl Foerster
Molinia
Festuca

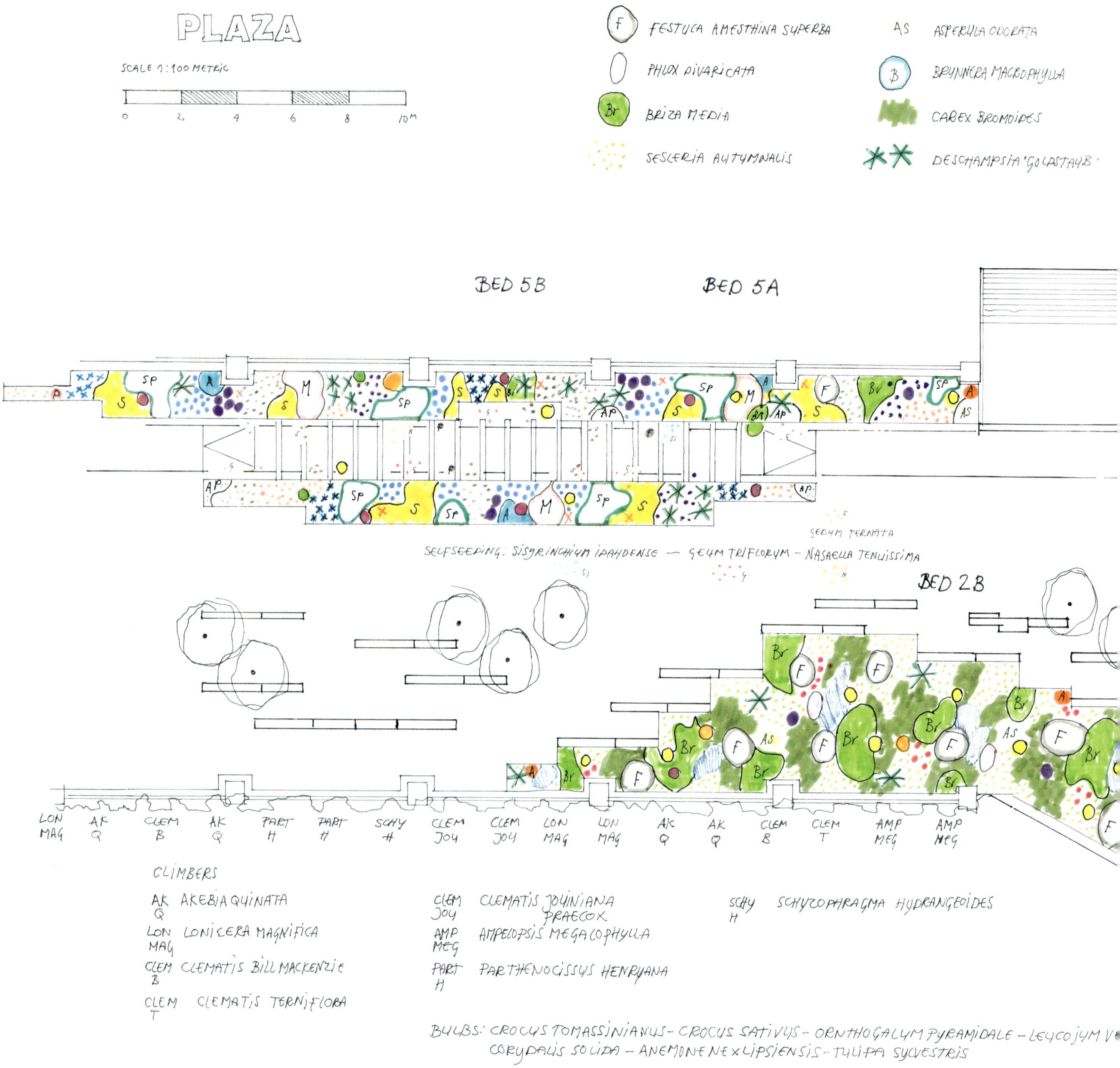

Final sketches for the Rail Yards Section, phase 1. This is a final rendering of a long process whereby different layers (woody plants, perennials, scatter plants) are built up on tracing paper. This was the first time that Piet used this technique, a measure of the complexity of this project, and of the need to ensure that all the layers could be read by visitors as they walked along.

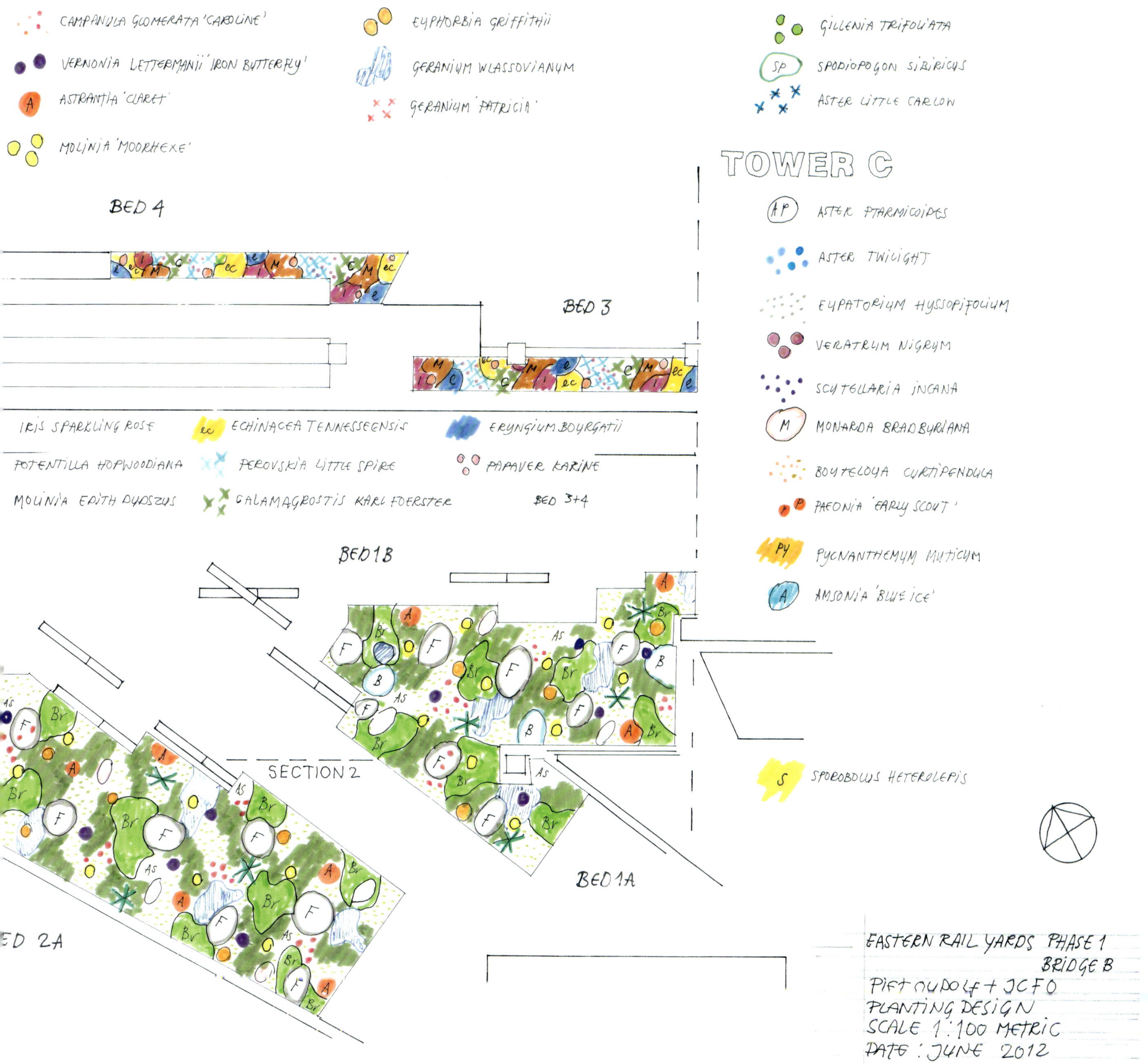
CAMPANULA GLOMERATA 'CAROLINE'
VERNONIA LETTERMANII 'IRON BUTTERFLY'
ASTRANTIA 'CLARET'
MOLINIA 'MOORHEXE'
EUPHORBIA GRIFFITHII
GERANIUM WLASSOVIANUM
GERANIUM 'PATRICIA'
GILLENIA TRIFOLIATA
SPODIOPOGON SIBIRICUS
ASTER LITTLE CARLOW
TOWER C
ASTER PTARMICOIDES
ASTER TWILIGHT
EUPATORIUM HYSSOPIFOLIUM
VERATRUM NIGRUM
SCUTELLARIA INCANA
MONARDA BRADBURIANA
BOUTELOUA CURTIPENDULA
PAEONIA 'EARLY SCOUT'
PYCNANTHEMUM MUTICUM
AMSONIA 'BLUE ICE'
SPOROBOLUS HETEROLEPIS
BED 4
BED 3
IRIS SPARKLING ROSE
ECHINACEA TENNESSEENSIS
ERYNGIUM BOURGATII
POTENTILLA HOPWOODIANA
PEROVSKIA LITTLE SPIRE
PAPAVER KARINE
MOLINIA EDITH DUDSZUS
CALAMAGROSTIS KARL FOERSTER
BED 3+4
BED 1B
SECTION 2
BED 1A
ED 2A
EASTERN RAIL YARDS PHASE 1
BRIDGE B
PIET OUDOLF + JCFO
PLANTING DESIGN
SCALE 1:100 METRIC
DATE : JUNE 2012

Previous pages: Aerial view of the Spur at 30th Street on the High Line, one of the few areas where there is more extensive planting, developed as a woodland of small-growing tree species as well as a space for public events. This was the last part of the original rail structure to be saved. Woodland on the High Line has a rich ground flora with a strongly naturalistic aesthetic (above). The sculpture, *Brick House* (2019) by Simone Leigh, is the inaugural commission for the High Line Plinth, which will feature a changing series of public artworks.

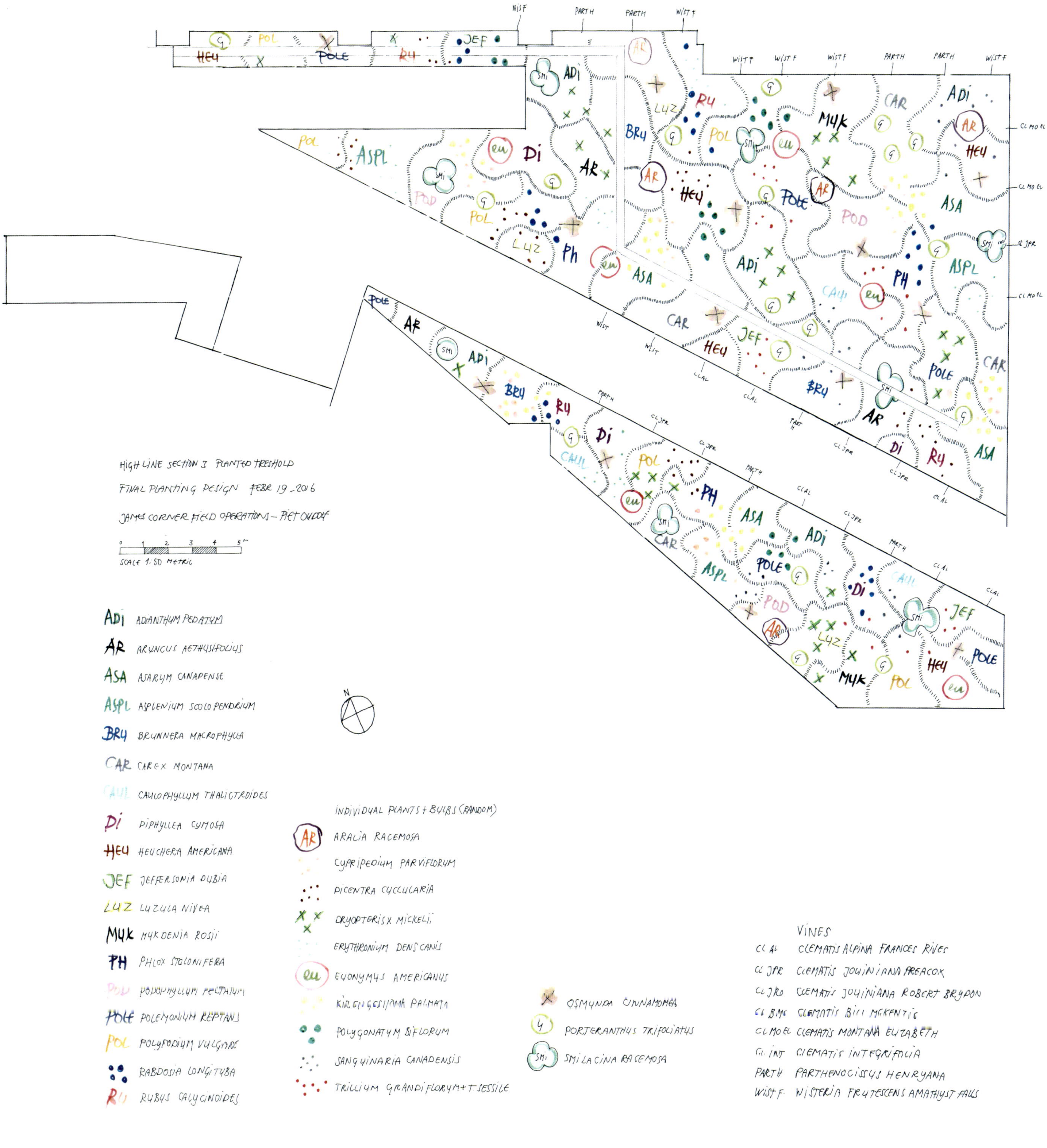

Plan for the Spur at 30th Street in phase 3; this is a woodland planting, where native northeastern US perennials (including bulbs, some grasses and ferns) form a ground layer beneath small tree species. A number of climbing plants, not necessarily native, including several clematis species and cultivars, are used to scramble up the steel exterior fencing.

An amphitheatre at the Spur and its accompanying piazza, planting for a small area of which is shown opposite.

Overleaf: The prairie-style shade-tolerant planting in the Chelsea Grasslands, which was part of phase 2, features *Aster* 'October Skies' and *Amsonia hubrichtii*, among other plants.

North
Monarda
THE SPUR PIAZZA
DESCHAMPSIA GOLDTAU
CAREX PENSYVANICA
DICTAMNUS ALBUS
ASTER LATERIFLORUS HORIZONTALIS
DRYOPTERIS ERYTHROSORA
EUPHORBIA BONFIRE
GENTIANA ANDREWSII
GERANIUM BROOKSIDE
GILLENIA TRIFOLIATA
LOBELIA SIPHILITICA
GERANIUM WLASSOVIANUM + ASTRANTIA ROMA
MONARDA BRADBURIANA
POLEMONIUM CAERULEUM
TRICYRTIS FORMOSANA
PERSICARIA FILIFORMIS
SMILACINA RACEMOSA
FORTHERGILLA GARDENII

UNDERLINE COFFEE

Collaboration: An Evolving Process

James Corner

Piet Oudolf is a master of his medium. He is a singular genius in terms of plants and planting design, the result of decades of hands-on experience making gardens. On the other hand, so much of his success has been born from creative collaboration with others, something Piet would be the first to acknowledge. His work alongside clients, architects, landscape architects, artists, plant nurseries, soil specialists, ecologists, gardeners, and installation and maintenance crews, among others, has led to extraordinarily productive exchanges wherein the whole far exceeds the sum of the parts.

As a master of his craft, Piet brings an assured confidence and unparalleled skill; as a creative collaborator, he brings an open mind and a penchant for learning and growing. This double act of stalwart reliability with adaptive responsiveness allows him to continuously evolve his craft, so that he can be likened to the very meadows and landscapes he creates: bold and confident, yet yielding and ever-changing. Piet, like his gardens, is both a solo act and a collective, a robust confidence and a fluid, open process – a dynamic ecology of imagination.

I have had the pleasure of working with Piet Oudolf on a number of projects, most notably New York's High Line (pp.82–95), London's South Park at Queen Elizabeth Olympic Park (pp.124–29) and a large private estate on Nantucket island, off Cape Cod, Massachusetts. In each case the design evolved collaboratively, with multiple voices and inputs, and was deeply informed by the specificity of site, environment and context.

For the High Line, I had an early idea that the walker should experience a sequence of changing landscapes, each a self-contained biotope reflecting local conditions of sun, wind, moisture, adjacency and vista. We saw certain areas as woodland, tall meadow, wet meadow, habitat preserve or hedgerow, for example, and these indicated certain vegetational features, such as hanging vines, oak bosque, sumac grove and flowering display. At the same time, I wanted the entire journey to feel unified and holistic, not segmented, so that one felt one was moving through gradients of texture and scale – a choreography of scenes and experiences. The whole idea was to amplify the original 'found landscape' of the railway bed, with wild mixtures of grasses, perennials and other plant combinations effectively self-adapting to local micro-climates and opportunistically growing a living ecosystem on an otherwise inert, inorganic structure of concrete and steel.

Piet's naturalistic approach to planting design was completely in sync with this vision, and we worked well together from the very beginning. His ideas of working predominantly with a dynamic 'matrix' planting that lends a sense of scale and unity, within which are scattered a rich variety of accents and focal points – varied by micro-climate and biotope – suited the vision of the High Line as a sequence of gardens along a journey; but he went further, with the sheer diversity of mixtures and combinations used throughout, deploying literally hundreds of species. Piet designed a remarkable sense of wildness, combined with apparently endless variation and diversity. The effect is exhilaratingly rich and profuse, with a stunning abundance of alternating textures, colours, scents and scales, each unfolding over time, with sequential blooms and displays emerging throughout the year.

At the beginning of the design process, we would send drawings, plans and images to Piet to communicate early structuring ideas and themes; he would mark them up, annotate them, send them back; we would further adapt, modify and discuss them with Piet; he would again revise, update and alter, and so the process continued through discussions and iterative drawings. Once the overall structure was set, Piet prepared more elaborate details, with species names and spacings, and we translated those notes into technical drawings for the construction specifications. At the same time, gardeners, soil specialists and maintenance managers reviewed and commented on the design, leading to a further tuning of the final drawing set.

Such collaborative back-and-forth continues during plant installation in most of Piet's projects, in which he spray-paints the planting plan on the ground

I'm interested in the life cycle of how plant material ages over the course of the year, and how it relates to the plants around it. Like a good marriage.

Piet Oudolf

itself and then arranges specific plants. Individual pots are placed on the painted ground, whereupon he makes various adjustments *in situ*, moving things around until the composition, mix and spacing are just how he wants them. Others may comment on and participate in this activity, with Piet accepting some ideas and rejecting others; he remains the creative designer painting his vision, while others around him will actually execute the final planting once he has approved the layout (see 'Planting the Vitra Garden', p.210). It is a dynamic process, part knowing and part instinct, but Piet is always engaged and hands-on.

This process has been enacted dozens of times over Piet's career. He has enjoyed many collaborations with very different types of site, client, designer and other experts, and he will confess that each is different, and each allows him to develop and evolve his own thinking and approach in interesting and important ways. In turn, his own extraordinary knowledge, passion, artistry and social disposition enrich those with whom he interacts.

Today, Piet visits his built gardens and meets the maintenance and gardening teams to discuss ongoing changes. Here, too, is a process of collaboration, a conversation between the designer and the caretakers of the garden, each discussing observations, ideas, conjectures, possibilities. There is nothing fixed in Piet's designs; they inevitably change and require active management, and as such they adapt and improve with time.

It is perhaps temporality that is Piet's most constructive collaborator. He learns through doing, observing and manipulating his gardens in time. He may act with a bold confidence, but he is also always listening, watching, observing and adapting as he goes along. He accepts open-endedness, the wildness of nature and the fact that his gardens are never finished – like him, the gardens actively collaborate, evolve and get better with time.

Leuvehoofd, Ichtushof and Westerkade

Rotterdam, The Netherlands
2010

Like many such urban port and industrial areas, the former quayside of Rotterdam has been redeveloped, with planners from the late 1980s onwards seeking to create an attractive, contemporary, pedestrian-friendly area featuring high-profile architecture and landscaping. The Erasmus Bridge over the Nieuwe Maas (New Quarter) is key here, creating a new link between the city centre and the southern port area. Piet has been involved in planting in three areas, which give a distinctive character to a location that could all too easily have become a stereotypical urban regeneration.

In the Ichtushof area, a garden sits over the roof of a metro station, its surroundings dominated by tall buildings. Piet's contribution is planting dominated by birch trees and grasses, with an extensive bulb planting for spring. Leuvehoofd is much more open, the eye being directed out towards the Erasmus Bridge by a river of *Deschampsia cespitosa* 'Goldschleier' grasses flowing across four planting beds. There is further planting for around 300 metres (330 yards) along the Westerkade (Western Quay), another planted intervention in the former waterfront area with a series of geometric planting beds, each one relatively small (some only 100 square metres/120 sq. yards) and each with a simplified planting scheme, many with an unusual integration of shrubs with perennials and grasses.

Leuvehoofd, with its distinctive river of grass running across the image.

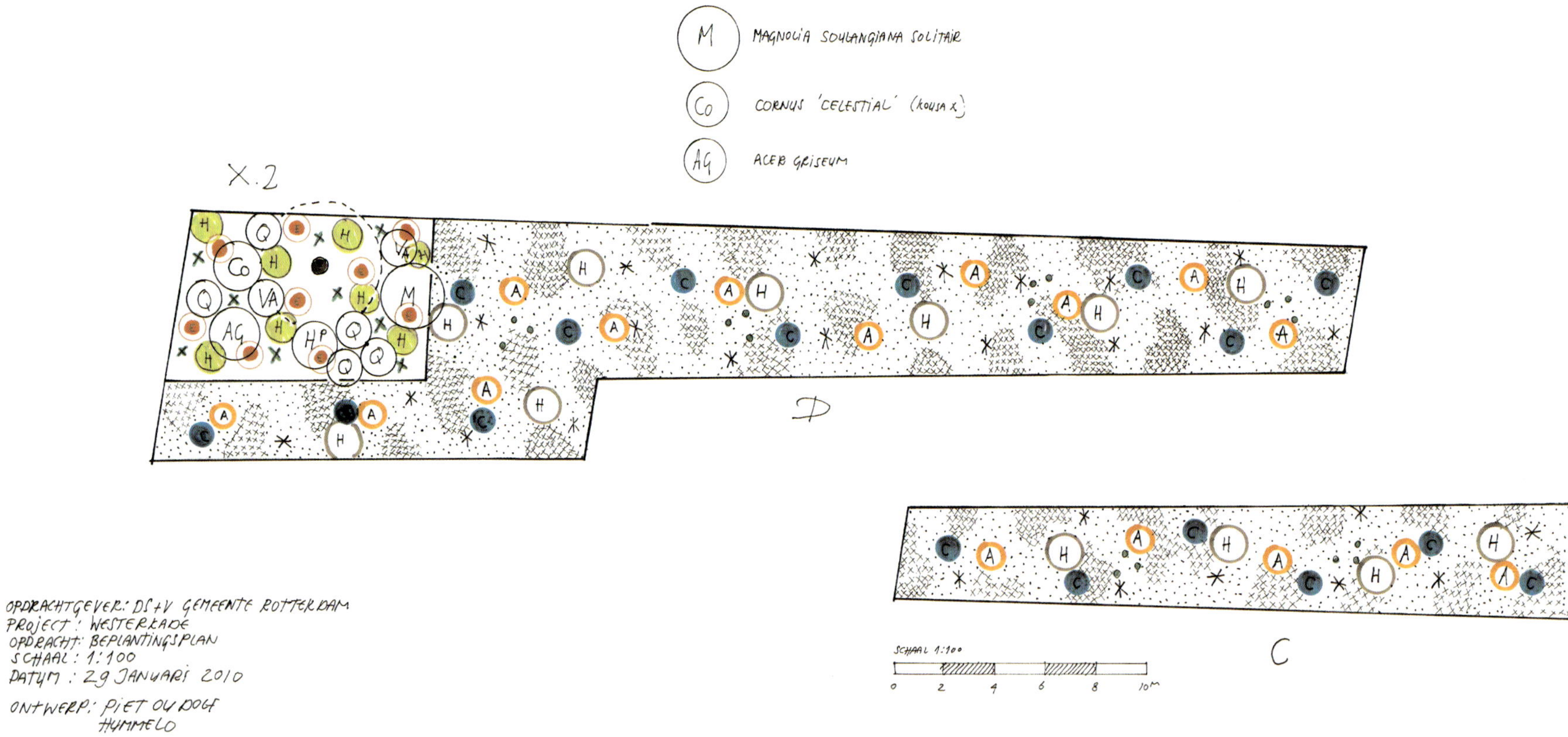

Top: A planting plan for five of the Westerkade beds. *Above, left to right*: A view of Westerkade and two views of Leuvehoofd; Westerkade is also shown on the following pages. A matrix style (left) is used in a very high proportion of the area of these plantings, including the lush-looking grass *Spodiopogon sibiricus*.

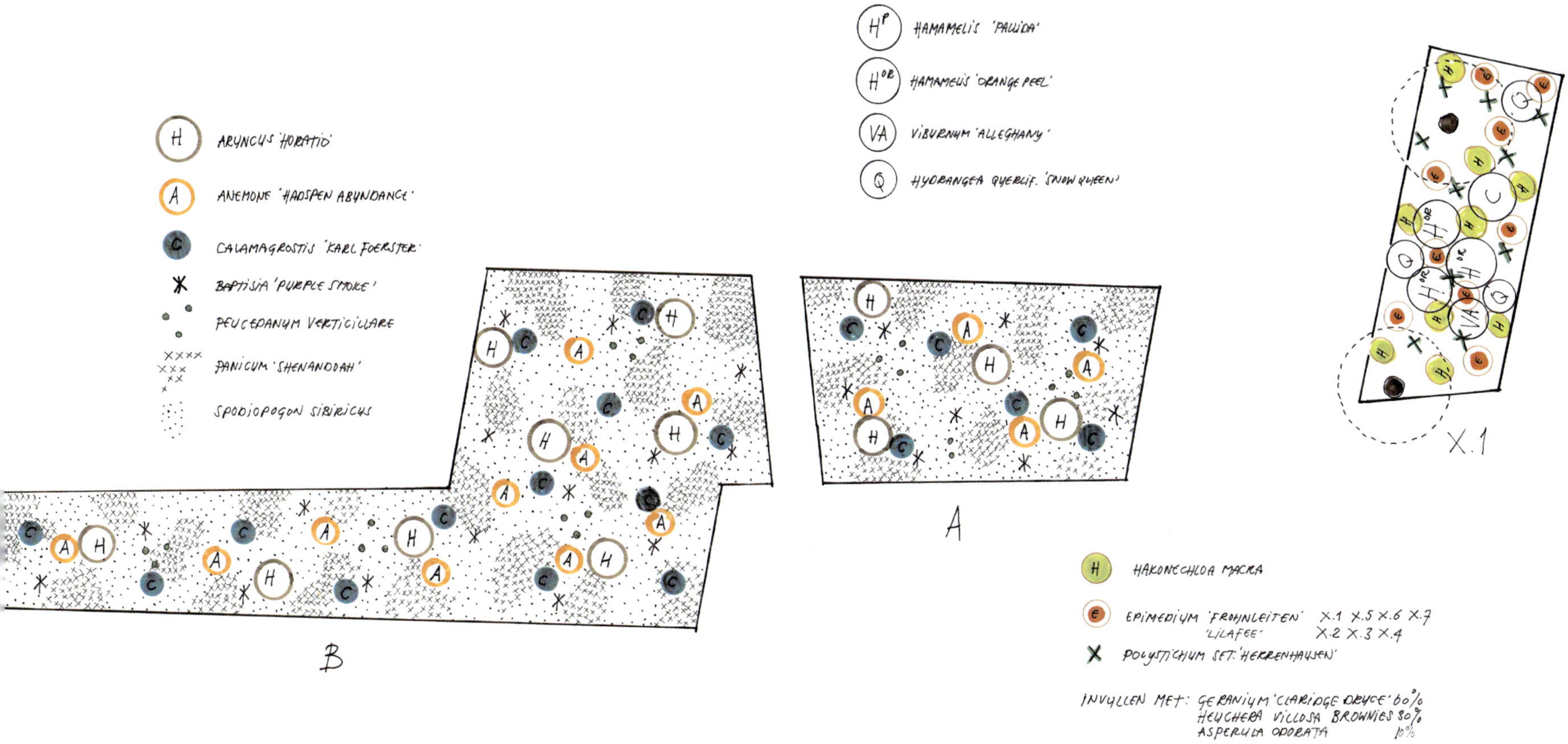
HAMAMELIS 'PALLIDA'
HAMAMELIS 'ORANGE PEEL'
VIBURNUM 'ALLEGHANY'
HYDRANGEA QUERCIF. 'SNOW QUEEN'
ARUNCUS 'HORATIO'
ANEMONE 'HADSPEN ABUNDANCE'
CALAMAGROSTIS 'KARL FOERSTER'
BAPTISIA 'PURPLE SMOKE'
PEUCEDANUM VERTICILLARE
PANICUM 'SHENANDOAH'
SPODIOPOGON SIBIRICUS
X.1
A
B
HAKONECHLOA MACRA
EPIMEDIUM 'FROHNLEITEN' X.1 X.5 X.6 X.7
'LILAFEE' X.2 X.3 X.4
POLYSTICHUM SET. 'HERRENHAUSEN'
INVULLEN MET: GERANIUM 'CLARIDGE DRUCE' 60%
HEUCHERA VILLOSA BROWNIES 30%
ASPERULA ODORATA 10%

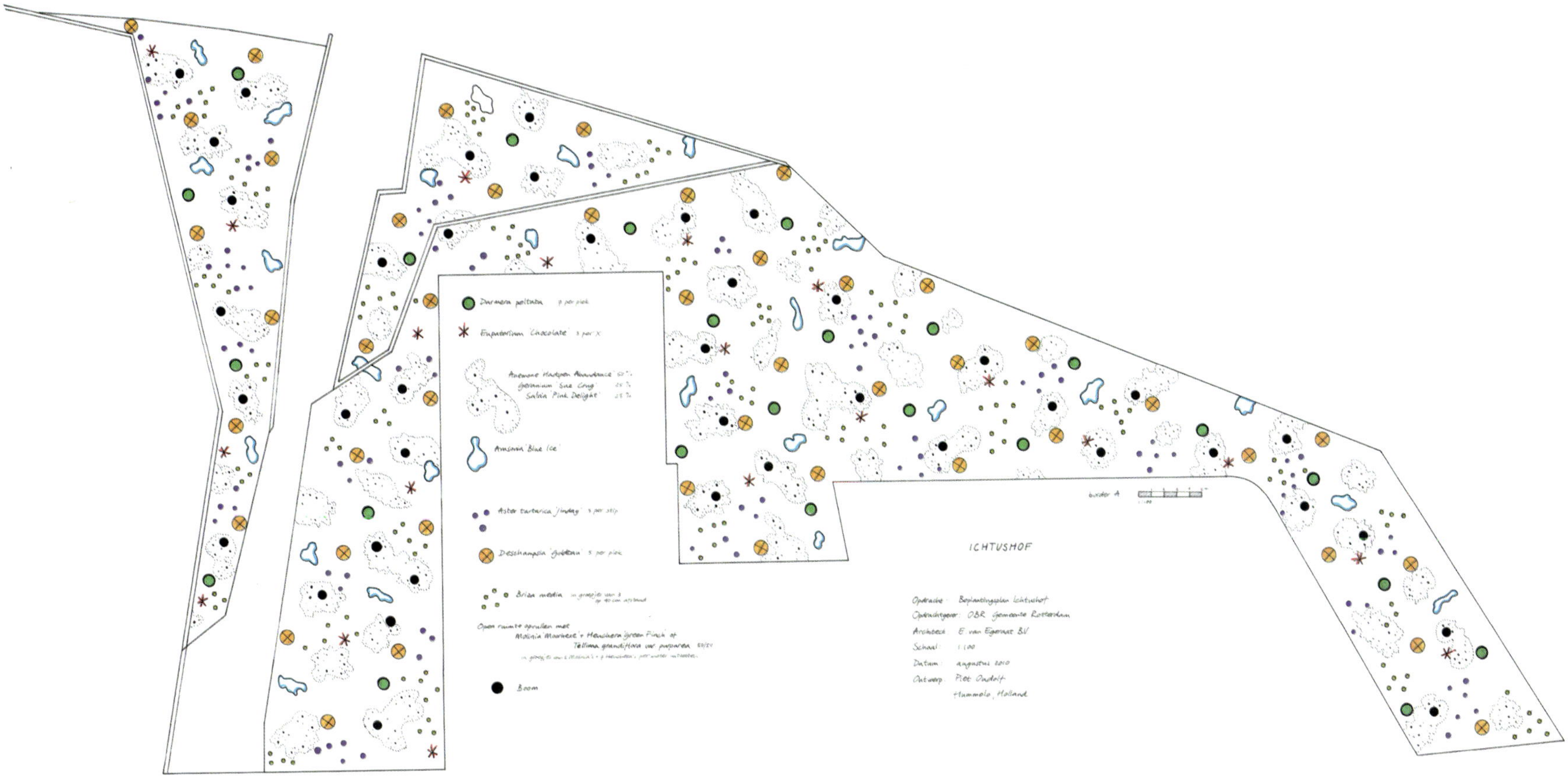

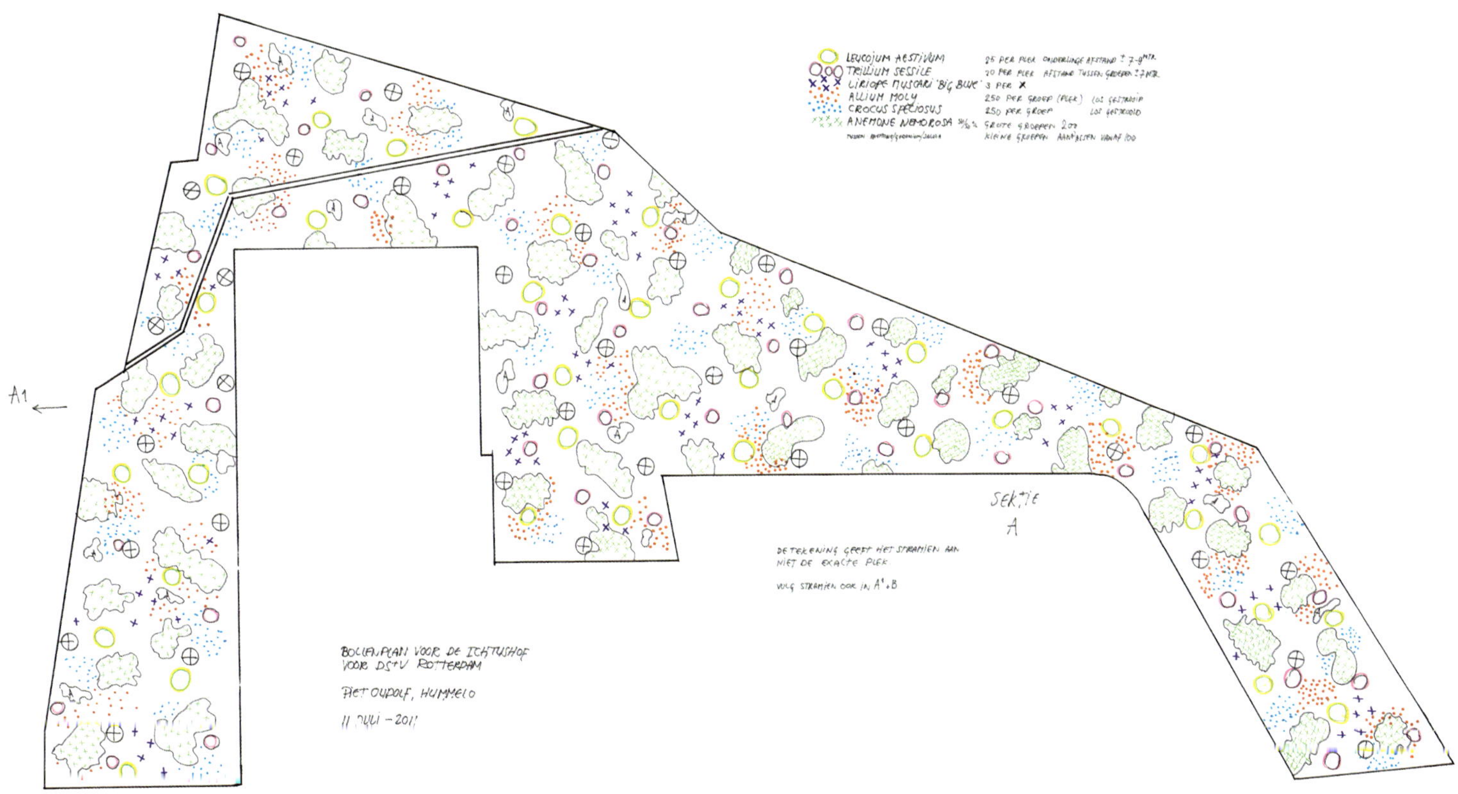

Ichtushof, near the entrance to the metro station. Perennial and tree placement plan above, and the bulb plan below. The planting style here uses relatively few species compared to most of Piet's designs.

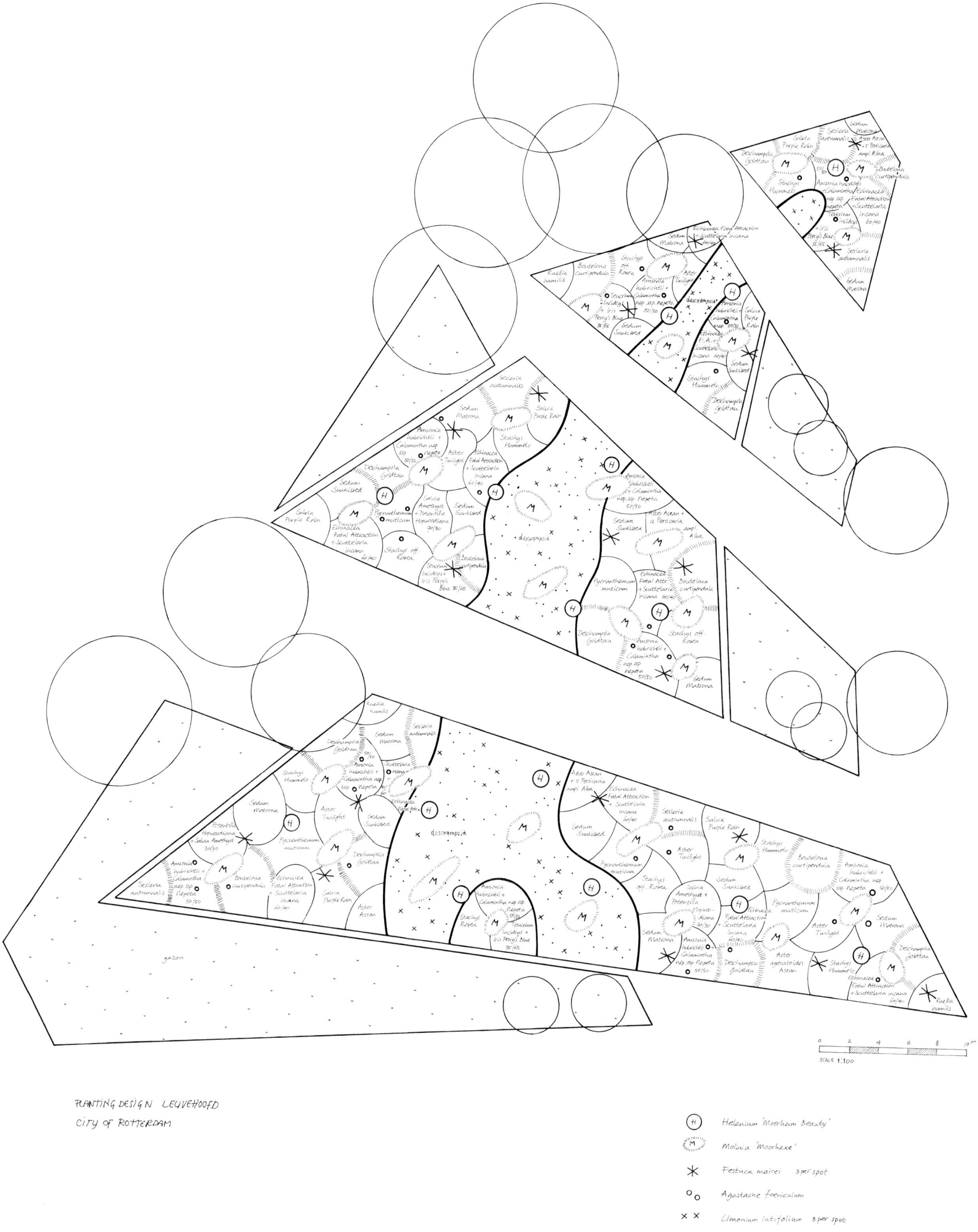
PLANTING DESIGN LEUVEHOOFD
CITY OF ROTTERDAM
Helenium 'Moerheim Beauty'
Molinia 'Moorhexe'
Festuca mairei 3 per spot
Agastache foeniculum
Limonium latifolium 3 per spot
Sedum Sunkissed verder invullen met Deschampsia goldschleier
SCALE 1:100
gazon
deschampsia

Leuvehoofd, planting plan and view. The central 'river' of grass, *Deschampsia cespitosa* 'Goldschleier', is a striking and dynamic feature.

'Il Giardino delle Vergini' at the Venice Biennale

Venice, Italy
2010

This planting was created in 2010, in collaboration with Japanese architect Kazuyo Sejima of SANAA, for the renowned international art event that takes place among the remains of the historic naval depot of the Venice Arsenale. Plane trees were planted here in a previous era, but there had been no other attempt at landscaping the site. The herbaceous nature of Piet's planting evokes the varied history of the site, long abandoned and overrun with spontaneous vegetation. The usual block planting style is followed, but with scattered structural plants.

The commissioning of Piet by an art and architecture festival inspired the art world, which has traditionally expressed little interest in the horticultural aspects of garden or landscape design, to pay him more attention. The following year he was commissioned by the Serpentine Gallery in London to complement a pavilion by Swiss architect Peter Zumthor. The result was an enclosed, contemplative *hortus conclusus*, for the 2011 summer-long exhibition at the gallery – one of the few temporary projects Piet has undertaken (see p.250). The Garden of the Virgins in Venice also attracted the interest of Switzerland-based art dealers Hauser & Wirth, who commissioned a masterplan for gardens at their new gallery at Durslade Farm in Somerset in southwestern England (pp.130–49).

Asters and Japanese anemones signal the turning of summer into autumn. Many of the species here have already developed seed heads by this time.

TOWER
TANK
RYE
RYE
A
C
TREE
GARDEN AND PLANTING DESIGN (FINAL)
PIET OUDOLF 2010
EUPATORIUM MAC ATROPURPUREUM
MOLINIA TRANSPARENT
HELENIUM RUBINZWERG
ECHINOPS TAPLOW BLUE
ASTER N.A VIOLETTA
ANEMONE ROBUSTISSIMA
PERSICARIA ORANGE FIELD
AGASTACHE BLUE FORTUNE
ACONITUM ARENDSII
PENNISETUM VIRIDESCENS

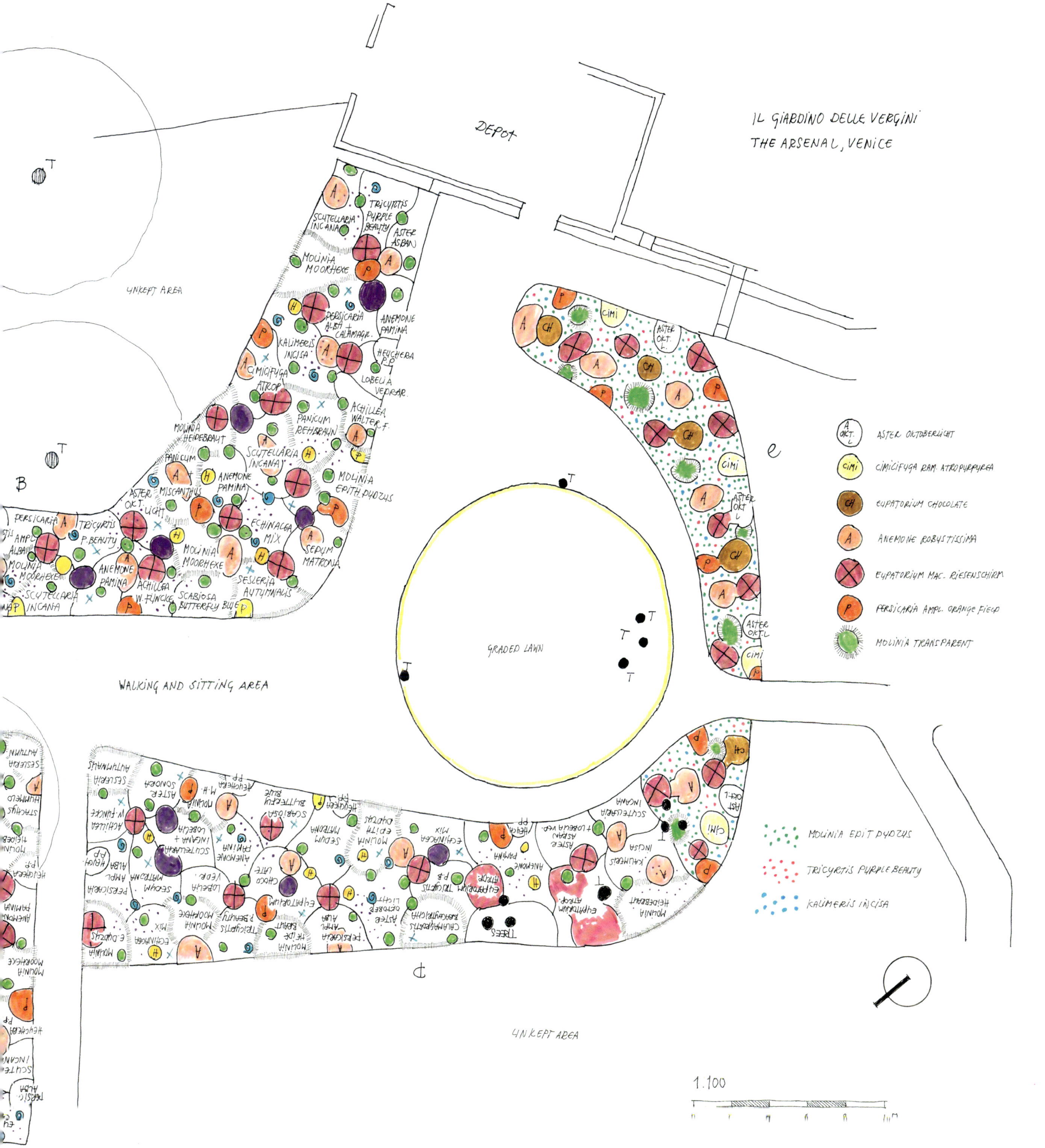

The planting plan for the Garden of the Virgins illustrates very clearly a combination of block planting and an almost random scattering of plant species that make an impact with a long season of structure, such as grasses and robust, often taller perennials.

Existing plane trees bring welcome shade and context, with woodland species beneath that can compete with established tree roots. A lawn contrasts with the deliberately naturalistic look of the perennial beds.

Maximilian Park
Hamm, Germany
2011–18

First completed in 2011 but with the addition of a new border planting in 2018, this public park in northwestern Germany is one of surprisingly few Oudolf projects in that country. The site was developed for one of the regional garden shows (*Landesgartenschauen*) that are used so successfully in Germany to enable the regeneration of former industrial or other derelict sites – this one stands on a former coal mine. Having been constructed in 1984, the improvements made for the garden show were in need of some contemporary additions, hence this commission by the park's director.

This planting can be seen as typical of a style Piet has developed for larger public projects – such as De Vlinderhof in the Netherlands (see pp.150–55) and Trentham in the English Midlands – in which a roughly rectangular area is divided into organic shapes with paths running through, forming a network without any obvious privileging of one area over another or sense of direction or progression. Block-planted perennials are interspersed with an almost random distribution of grasses or perennials that have a long season of structural interest; a variant on the idea of 'scatter plants', these act as rhythmic elements.

Areas of shade, created by the need to work around existing trees, are home to perennials chosen for their tolerance of reduced light. Many, such as *Kirengeshoma palmata* (lower right), are notable for their distinctive foliage.

Maxi parks
TRICYRTIS
Aster Twilight
Kirengeshoma
Hakonechloa
Sanguisorba canadensis
Acanthus mollis
Brunnera
Clematis Cassandra

An early sketch for the border added in 2018 at Maximilian Park. The planting creates a link between an entrance planting and the main perennial area. The right of the two oval areas is a shaded bed beneath an oak, where woodland species predominate.

The planting here illustrates the strength of Piet's autumn look, where contrasts in form and visual texture between seedheads are the dominant attraction. This is one of the most important and innovative features of his work as a planting designer.

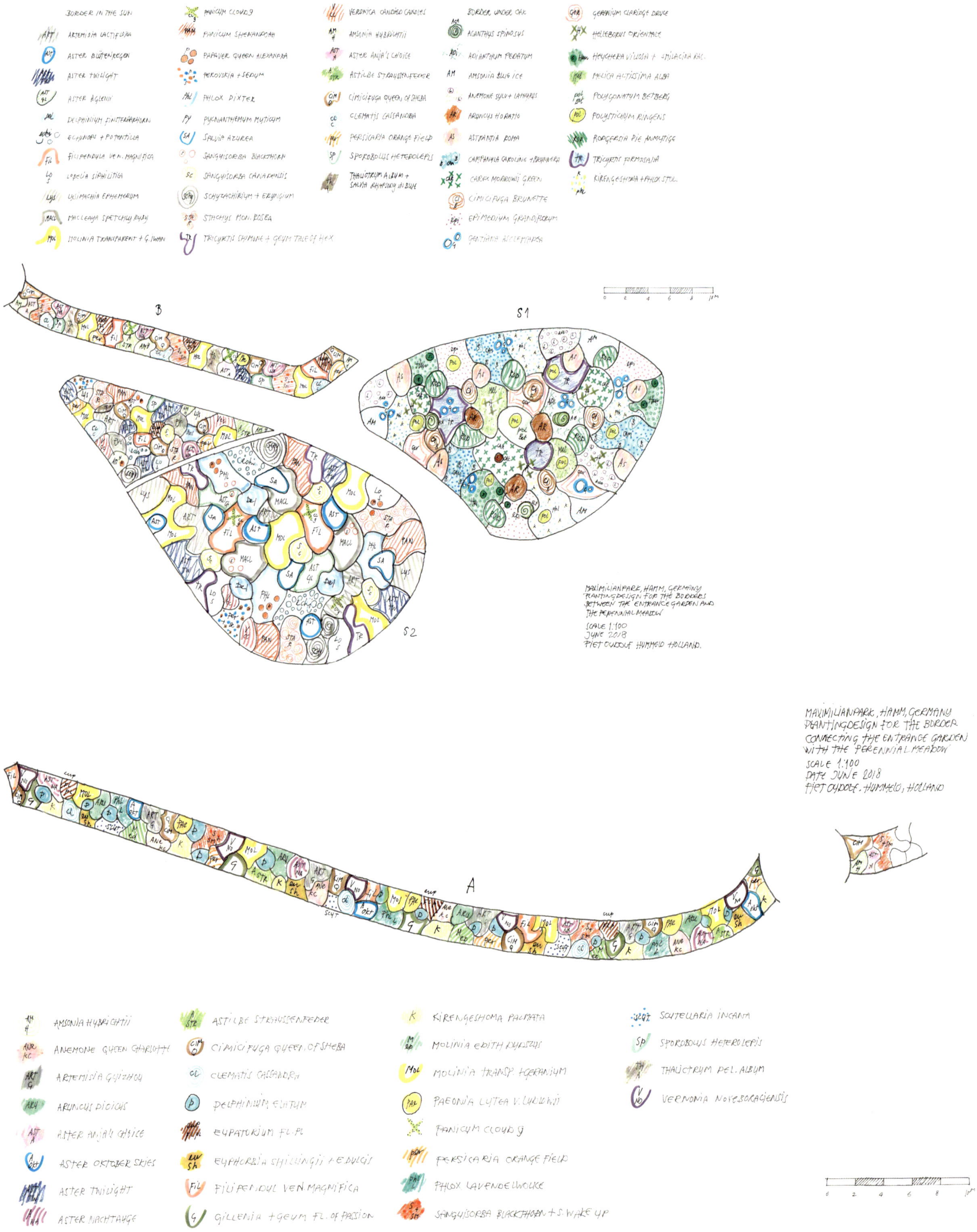
MAXIMILIANPARK, HAMM, GERMANY
PLANTINGDESIGN FOR THE BORDERS
BETWEEN THE ENTRANCE GARDEN AND
THE PERENNIAL MEADOW
SCALE 1:100
JUNE 2018
PIET OUDOLF HUMMELO HOLLAND.
MAXIMILIANPARK, HAMM, GERMANY
PLANTINGDESIGN FOR THE BORDER
CONNECTING THE ENTRANCE GARDEN
WITH THE PERENNIAL MEADOW
SCALE 1:100
DATE JUNE 2018
PIET OUDOLF, HUMMELO, HOLLAND

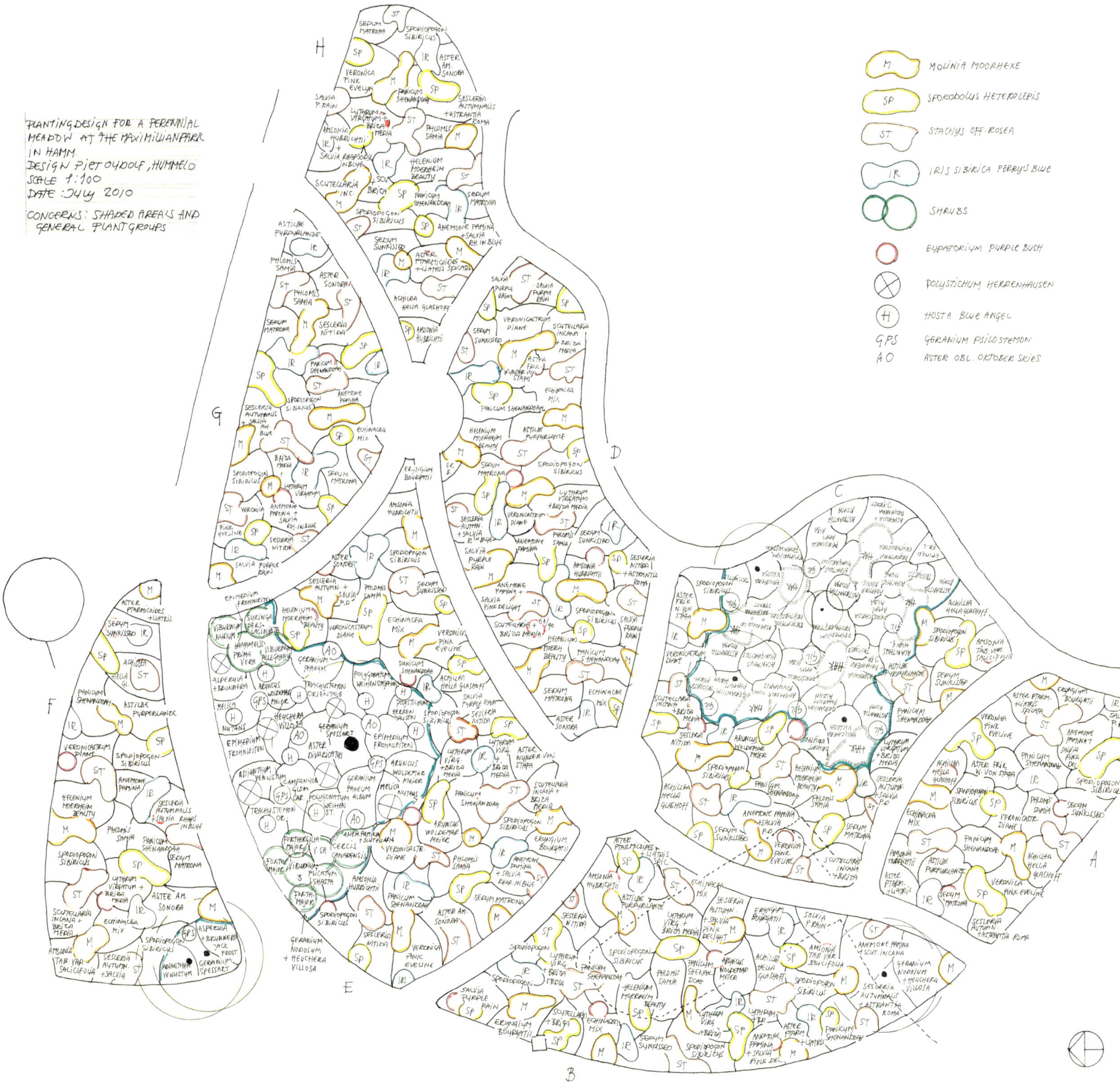

Plans for the 2018 Maximilian Park planting (opposite), which extended the original design. Above, a planting plan for the 2010 installation. Rhythm across the planting is achieved through repetition of certain key species, which are indicated in the legend at upper right. *Overleaf*: *Aster novae-angliae* 'Nachtauge' (foreground) in early autumn. This is a view of the 2018 design area, which connects the entrance planting and the Perennial Meadow of 2010.

Queen Elizabeth Olympic Park
London, UK
2012–14

The 2012 London Olympics was an opportunity for some of Britain's most innovative planting designers to show off their work. The site, in east London, was then redeveloped as a public park over the next two years, with planting by Piet as part of this next layer. The Pleasure Gardens is an area that was paved for the Olympics as a central piazza, then was extensively changed for the park. The central concept is that this area should host a multiplicity of activities – play, music, family picnics, theatre – all of which require a degree of separation from each other. This differentiation of space was achieved with ribbons of planting, designed in collaboration with James Corner Field Operations, with whom Piet worked on New York's High Line as well as a large private garden on Nantucket island, off the coast of Cape Cod, in Massachusetts.

Shallow planting beds create particular challenges in design, as everything is on show all the time. Much of the planting is four metres wide, often with three plant blocks from side to side, two on the outside and one in the middle. The planting is perhaps more colourful than much of Piet's work, and it is all block planting; there are a lot of varieties of late-flowering genera, such as *Symphyotrichum* and *Eutrochium* (still much more commonly known by their former names, *Aster* and *Eupatorium*), which have been classics of the late season border for over a century. The emphasis on late summer is here entirely appropriate, for it is during mid- to late summer that this area is most intensively used by the diverse community that lives locally.

The heights of the plants in the relatively narrow ribbons of planting in the Queen Elizabeth Olympic Park during late summer help to create discrete areas for varied public uses of the park in this very popular public space. Here, dark pink-purple *Eupatorium maculatum* 'Riesenschirm' and mauve *Phlox* 'Great Dixter' make a strong impact.

Design sketch; here Piet is orchestrating the number and distribution of perennials for the narrow ribbons of planting in the Olympic Park.

Bands of colourful planting are illustrated in a final planting plan and legend, top and right; it is worth noting a few examples of where Piet has used blocks of a combination of two species, for example *Veratrum nigrum* and *Stachys officinalis*, the latter filling in the ground space between plants of the strongly upright *Veratrum*. Above, a detail of the planting. In the foreground the pink *Lythrum* 'Swirl' is interplanted with *Parthenium integrifolium*. The grass *Calamagrostis* × *acutiflora* 'Karl Foerster' is in the background and has proved a key species, not just for Piet but for the whole New Perennial movement.

KEY PERENNIALS (TALL GRASS/PERENNIAL AREA)

- ACHILLEA FIL. 'PARKER'S VAR.'
- ALTHAEA 'PARK RONDELL'
- ANEMONE TOM 'ROBUSTISSIMA'
- ARUNCUS 'HORATIO'
- ARUNCUS DIOICUS 'ZWEIWELTENKIND'
- ASTER 'LITTLE CARLOW'
- ASTER 'TWILIGHT'
- ASTER LAT. 'HORIZONTALIS'
- ASTER N-A 'VIOLETTA'
- ASTER TART. 'JINDAY'
- ASTER UMBELLATUS
- ASTRANTIA 'ROMA'
- BAPTISIA AUSTRALIS
- BAPTISIA 'LEUCANTHA'
- CIMICIFUGA 'PRITCHARD'S GIANT' + THALICTRUM ROCHEBRUNEANUM
- CIMICIFUGA 'QUEEN OF SHEBA' + KIRENGESHOMA PALMATA 70%-30%
- CRAMBE CORDIFOLIA
- DATISCA CANNABINA
- DELPHINIUM ELATUM
- ECHINACEA PALLIDA
- ECHINOPS BANNATICUS
- EUPATORIUM MAC 'SNOWBALL'
- EUPATORIUM MAC. 'RIESENSCHIRM'
- EUPATORIUM RUG. 'MUSK'
- FILIPENDULA RUBRA 'VENUSTA MAGNIFICA'
- GERANIUM 'CLARIDGE DRUCE' + GERANIUM 'PATRICIA' 80%-20%
- GERANIUM SPINNERS
- GILLENIA TRIFOLIATA
- HELENIUM 'LOYSDER WIECK'
- HELENIUM 'SAHIN'S EARLY FLOWER'
- HEMEROCALLIS CITRINA XOCHROLEUCA
- INULA MAGNIFICA 'SONNENSTRAHL'
- IRIS SIBIRICA 'PERRY'S BLUE' + 'CAESAR'S BROTHER' 25%-75%
- LOBELIA VEDRARIENSIS + MONARDA 'SCORPION' 65%-35%
- LYTHRUM 'SWIRL' + PARTHENIUM INTEGRIFOLIUM 65%-35%
- LYCHNIS CHALCEDONICA 'ROSEA' + LYSIMACHIA EPHEMERUM 30%-70%
- PAPAVER 'SCARLET O'HARA' - KARINA
- PENSTEMON 'HUSKER'S RED' + AMSONIA HUBRICHTII 80%-15%
- PENSTEMON 'HUSKER'S RED' + ASCLEPIAS INCARNATA 80%-20%
- PERSICARIA AMPLEXICAULE 'ALBA'
- PERSICARIA AMPL. 'ORANGE FIELD'
- PHLOMIS TUBEROSA 'LITTLE AMAZONE'
- PHLOX PANICULATA 'GREAT DIXTER'
- RUDBECKIA SUBT. 'HENRY EILERS'
- SANGUISORBA 'TANNA'
- SANGUISORBA CANADENSIS
- SANGUISORBA TENUIFOLIA 'PURPUREA'
- SEDUM 'SUNKISSED'
- SELINUM WALLICHIANUM
- SOLIDASTER 'LEMORE'
- THALICTRUM 'ELIN' / THALICTRUM ILLUMINATOR
- THALICTRUM POLYGAMUM + PHLOX DÜSTERLOHE 70%-30%
- VERATRUM NIGRUM + STACHYS OFF. 'HUMMELO' 40%-60%
- VERNONIA NOVEBORACENSIS
- VERONICA LONGIFOLIA 'BLAURIESIN'
- VERONICASTRUM 'ADORATION'
- VERONICASTRUM 'ERICA'
- VERONICASTRUM 'DIANE'

KEY GRASSES (TALL GRASS/PERENNIAL AREA)

- CALAMAGROSTIS BRACHYTRICHA
- CALAMAGROSTIS ACUTIFL. 'KARL FOERSTER'
- DESCHAMPSIA 'GOLDTAU'
- MISCANTHUS 'FERNER OSTEN'
- MOLINIA 'HEIDEBRAUT'
- MOLINIA 'TRANSPARENT'
- PANICUM VIRGATUM 'CLOUD 9'
- PANICUM VIRG. 'DALLAS BLUES'
- PANICUM VIRG. 'NORTHWIND'
- PANICUM VIRG. 'SHENANDOAH'
- PENNISETUM VIRIDESCENS
- SORGHASTRUM 'SIOUX BLUE'
- SPODIOPOGON SIBIRICUS
- STIPA GIGANTEA

Oudolf Field at Hauser & Wirth Somerset
Durslade Farm, Somerset, UK
2012–14

Superficially similar to several other plantings of similar size and overall shape, the Oudolf Field, commissioned by Iwan and Manuela Wirth for their international art gallery Hauser & Wirth in Somerset, southwest England, in fact incorporates a range of disperate elements. One is a powerful sense of progression; there is a firm axis from the rear of the gallery towards an architectural structure at the far end – the Radić Pavilion, used for events, designed by Chilean architect Smiljan Radić. The wide central axial path was designed to accommodate people attending functions. Some shallow gradient grass mounds break up the width of the path; they are another example of the circular shapes that crop up in Piet's work. There are a variety of planting styles: the edges are quite a traditional combination of perennial groups (indicated on the plan with simple black and white), while the various planting beds use a block-planting style with some scatter plants. There is a response to environment, with damp soil at the gallery end being communicated with a range of wetland species, and a pond.

Three beds use a *Sporobolus* matrix, *Sporobolus heterolepis* being one of the most successful species for this planting style, as there seems to be a high level of co-existence possible between its tight tussocks and many perennials, quite apart from the visual harmony created by the contrasting of fine flower and seedheads with more defined perennial forms. Elsewhere, the Cloister Courtyard, a distinct space between several of the former farm buildings, now converted to gallery space, elegantly combines a limited number of species for a cool and tranquil feel. Enclosure here is everything, whereas with the Oudolf Field the eye is frequently pulled away into the wider view.

This view is from the middle of the Oudolf Field back towards the gallery. Overleaf, we look towards the Radić Pavilion. The repeated drifts of perennials are the core of Piet's design and planting style and make the most impact from late summer (when these pictures were taken) onwards. The importance of gently contrasting plant textures can also be appreciated here, enhanced by the block planting style. In the foreground, the striking seedheads of *Echinacea pallida* are notable for their fine post-flowering interest.

+45 Chaerophyllum hirsutum roseum
camassia wordt niet van de oppvl. afgetrokken
GALLERY
POND
PLANTING DESIGN FOR THE PERENNIAL MEADOW
AT DURSLADE FARM
COMMISSIONED BY IWAN AND MANUELA WIRTH
DATE: SEPTEMBER 2012
SCALE 1:100
DESIGN: PIET OUDOLF, HUMMELO, HOLLAND
1:100 METRIC

an of the Oudolf Field, and accompanying xplanatory legends. The outer borders are hown in black and white, a legacy of both a revious (and conventional) style of notation nd an older planting style – the simple juxtapo-tion of perennials, in this case mostly taller rieties. The central planting beds illustrate Piet's typically more complex planting. Two matrix plantings should be noted: one, the wet area in front of the pond, uses *Briza media* 'Limouzi' and *Sesleria* 'Greenlee Hybrid', and in the centre right are the three beds planted with the grass *Sporobolus heterolepis* as the ground layer (see pp.130–31).

AREAS 1,2,3,4

- ACHILLEA 'HELLA GLASHOFF'
- ALLIUM 'SUMMER BEAUTY'
- ANEMONE X HYBRIDA 'HONORINE JOBERT'
- ANEMONE X HYBRIDA 'PAMINA'
- ASTER 'LITTLE CARLOW'
- ASTER AMELLUS 'SONORA'
- CALAMINTHA NEPETA SSP. NEPETA
- CRAMBE CORDIFOLIA
- DORYCNIUM HIRSUTUM
- ECHINACEA PURPUREA 'FATAL ATTRACTION' + PYCNANTHEMUM MUTICUM 60% + 40%
- ECHINOPS BANNATICUS
- ERYNGIUM ALPINUM
- GENTIANA ASCLEPIADEA + THALICTRUM DELAVAYI
- GERANIUM 'ROZANNE'
- GILLENIA TRIFOLIATA
- HELENIUM 'MOERHEIM BEAUTY'
- KNAUTIA MACEDONICA + ORIGANUM 'HOPLEYS'
- LIATRIS SPICATA + LIMONIUM LATIFOLIUM
- LOBELIA VEDRARIENSIS
- LYCHNIS CHALCEDONICA
- LYSIMACHIA EPHEMERUM
- MOLINIA CAERULEA 'EDITH DUDSZUS'
- MONARDA BRADBURIANA + NEPETA GOVANIANA 80% + 20%
- ORIGANUM 'HOPLEYS' + KNAUTIA MACEDONICA
- PANICUM 'SHENANDOAH'
- PAPAVER ORIENTALE 'KARINE'
- PENSTEMON DIGITALIS 'HUSKER RED'
- PHLOMIS RUSSELIANA
- SCUTELLARIA INCANA
- SEDUM 'MATRONA'
- SELINUM WALLICHIANUM
- SERRATULA SEOANEI
- SESLERIA AUTUMNALIS
- SPOROBOLUS HETEROLEPSIS
- STACHYS OFFICINALIS 'ROSEA'
- SUCCISA PRATENSIS
- THALICTRUM DELAVAYI + GENTIANA ASCLEPIADEA
- VERONICASTRUM VIRGINICUM 'ERICA'

AREA 5,6,7

- ACHILLEA 'CREDO'
- AMSONIA HUBRICHTII
- ASTER LATERIFLORUS 'HORIZONTALIS'
- ASTER MACROPHYLLUS 'TWILIGHT'
- ASTER TATARICUS 'JINDAI'
- BOUTELOUA CURTIPENDULA
- BRIZA MEDIA 'LIMOUZI' + MOLINIA 'MOORHEXE' 60% + 40%
- CHAEROPHYLLUM HIRSUTUM 'ROSEUM'
- ECHINACEA PALLIDA + STIPA TENUISSIMA 70% + 30%
- ERYNGIUM ALPINUM
- ERYNGIUM YUCCIFOLIUM
- GERANIUM 'PATRICIA'
- HELENIUM 'MOERHEIM BEAUTY'
- LYSIMACHIA EPHEMERUM
- LYTHRUM SALICARIA 'BLUSH'
- MOLINIA 'MOORHEXE' + CALAMINTHA NEPETA SSP. NEPETA
- NEPETA SUBSESSILIS
- PANICUM 'SHENANDOAH'
- PAPAVER ORIENTALIS 'SCARLET O'HARA'
- PENNISETUM VIRIDESCENS
- PENSTEMON 'HUSKER RED'
- PEROVSKIA
- PYCNANTHEMUM MUTICUM
- SALVIA 'DEAR ANJA'
- SALVIA VERTICILLATA 'PURPLE RAIN'
- SANGUISORBA 'RED BUTTON'
- SCABIOSA COLUMBARIA + LIMONIUM LATIFOLIUM
- SCUTELLARIA INCANA
- SEDUM 'MATRONA'
- SESLERIA AUTUMNALIS
- STACHYS 'HUMMELO'
- STIPA TENUISSIMA + ECHINACEA PALLIDA
- THALICTRUM DIPTEROCARPUM + GENTIANA ASCLEPIADEA
- VERONICASTRUM VIRGINICUM 'ERICA'

SPOROBOLUS MEADOW 1,2,3

- ACHILLEA 'FEUERLAND'
- AGASTACHE
- AMORPHA CANESCENS
- AMSONIA HUBRICHTII
- ASCLEPIAS INCARNATA
- ASTER TATARICUS 'JINDAI'
- DIANTHUS CARTHUSIANORUM
- ECHINACEA PALLIDA 'HULA DANCER'
- FESTUCA MAIREI
- IMPERATA CYLINDRICA
- IRIS CHRYSOGRAPHES 'BLACK FORM'
- LYTHRUM SALICARIA 'SWIRL'
- MOLINIA CAERULEA 'MOORHEXE' + LIMONIUM LATIFOLIUM
- MONARDA BRADBURIANA
- PERSICARIA AMPLEXICAULIS 'ALBA'
- RUELLIA HUMILIS
- SCABIOSA COLUMBARIA
- SEDUM 'CORAL REEVES'
- SPOROBOLUS HETEROLEPSIS

WET AREA

- ANEMONE 'WILD SWAN'
- ASTILBOIDES TABULARIS
- ASTRANTIA MAJOR 'ROMA'
- BRIZA MEDIA 'LIMOUZI'
- CAMASSIA LEICHTLINII
- CIMICIFUGA SIMPLEX 'JAMES COMPTON'
- CLEMATIS HERACLEIFOLIA 'CHINA PURPLE'
- DARMERA PELTATA
- GERANIUM 'SPINNERS'
- IRIS SIBIRICA 'PERRY'S BLUE'
- LOBELIA TUPA
- LYTHRUM SALICARIA 'BLUSH'
- MYRICA GALE
- PIMPINELLA MAJOR 'ROSEA'
- SERRATULA SEOANEI
- TRICYRTIS FORMOSANA

CORNERS 1 AND 2

- ANEMONE X HYBRIDA 'HONORINE JOBERT'
- ARUNCUS 'HORATIO'
- ASTER 'VIOLETTA'
- ASTER MACROPHYLLUS 'TWILIGHT'
- CIMICIFUGA SIMPLEX 'PRITCHARD'S GIANT'
- DESCHAMPSIA 'GOLDTAU'
- EUPATORIUM MACULATUM 'ATROPURPUREUM'
- FILIPENDULA RUBRA 'VENUSTA MAGNIFICA'
- HELENIUM 'KUPFERZWERG'
- MOLINIA 'TRANS PARENT'
- PERSICARIA AMPLEXICAULIS 'ORANGE FIELD'
- SANGUISORBA 'BLACKTHORN'
- SPODIOPOGON SIBIRICUS
- SUCCISA PRATENSIS
- SUCCISELLA INFLEXA
- THALICTRUM ROCHEBRUNNIANUM
- VERONICASTRUM VIRGINICUM 'ADORATION'

CORNERS 3 AND 4

- ANEMONE ROBUSTISSIMA
- ARUNCUS 'HORATIO'
- ASTER MACROPHYLLUS 'TWILIGHT'
- ASTER UMBELLATUS
- CIMICIFUGA SIMPLEX 'PRITCHARD'S GIANT'
- DATISCA CANNABINA
- DESCHAMPSIA CESPITOSA 'GOLDTAU'
- EUPATORIUM MACULATUM 'ATROPURPUREUM'
- FILIPENDULA RUBRA 'VENUSTA MAGNIFICA'
- HELENIUM 'LOYSDER WIECK'
- MOLINIA 'TRANSPARENT'
- PERSICARIA AMPLEXICAULIS 'FIRE DANCE'
- SANGUISORBA 'RED BUTTON'
- SANGUISORBA 'BURY COURT'
- SANGUISORBA CANADENSIS
- THALICTRUM 'ELIN'
- THALICTRUM ROCHEBRUNNIANUM
- VERONICASTRUM VIRGINICUM 'DIANE'

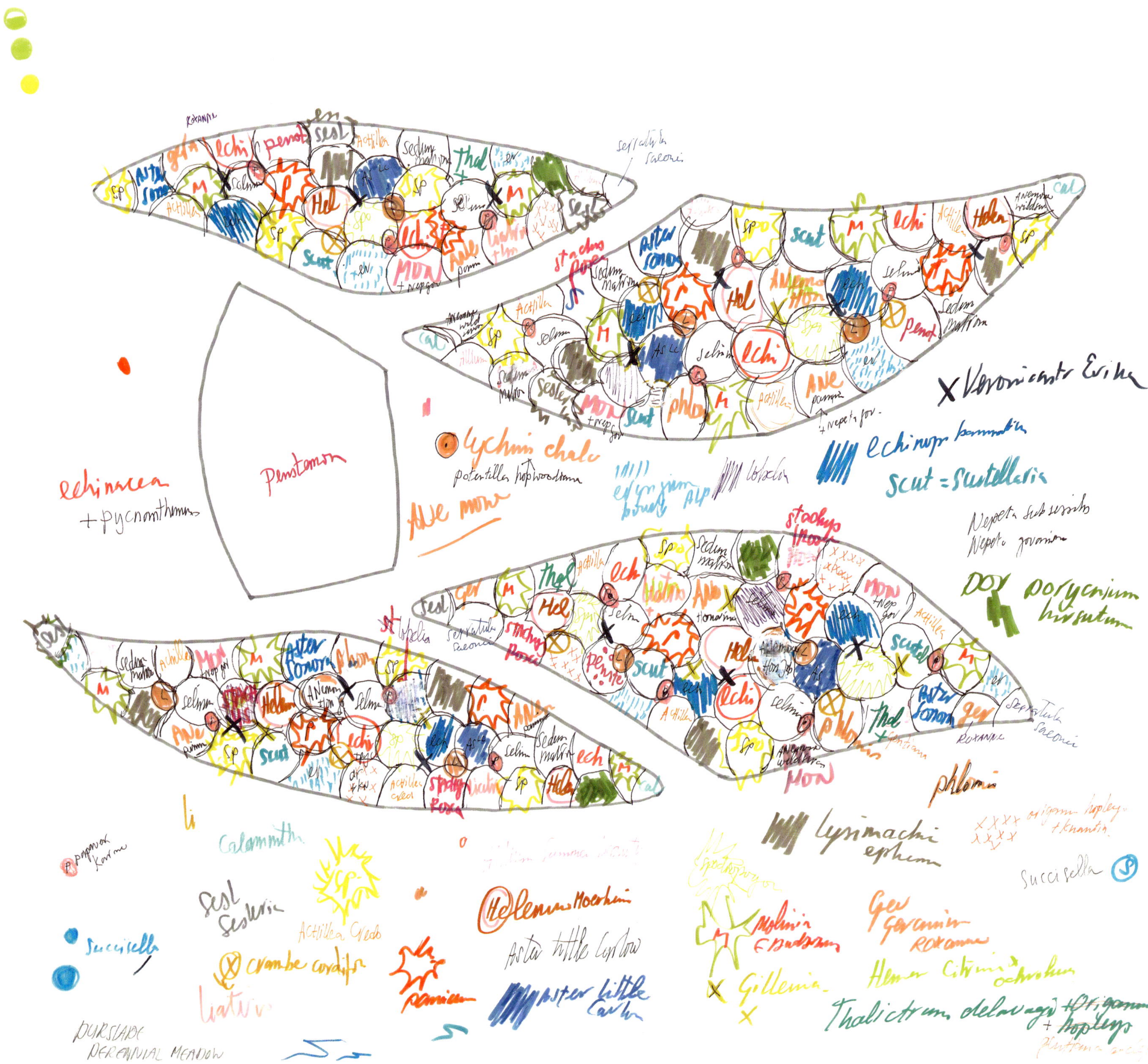

First draft sketches for Oudolf Field (also overleaf). These are used to experiment with basic ideas of key plants, their distribution and juxtaposition. There is no particular hierarchy or order in which Piet arranges plants, which means that there is more openness to new possibilities of arrangement.

Sporobolus
Imperata Cylind
Lythrum Swirl
Iris chrysographes
Hula dancer
Ruellia
Sedum Sunkissed
persicaria Alb.
Amorpha
Achillea Credo
Asclepias incarnata
Molinia Moorhexe
Limonium lat.
Dianthus Carthusianorum
Monarda bradburiana
Festuca Mairei
Amsonia hubrichtii
DURSLADE MEADOW
SPOROBOLUS

Typical block planting is interrupted by the very different texture of the *Sporobolus* matrix visible behind.

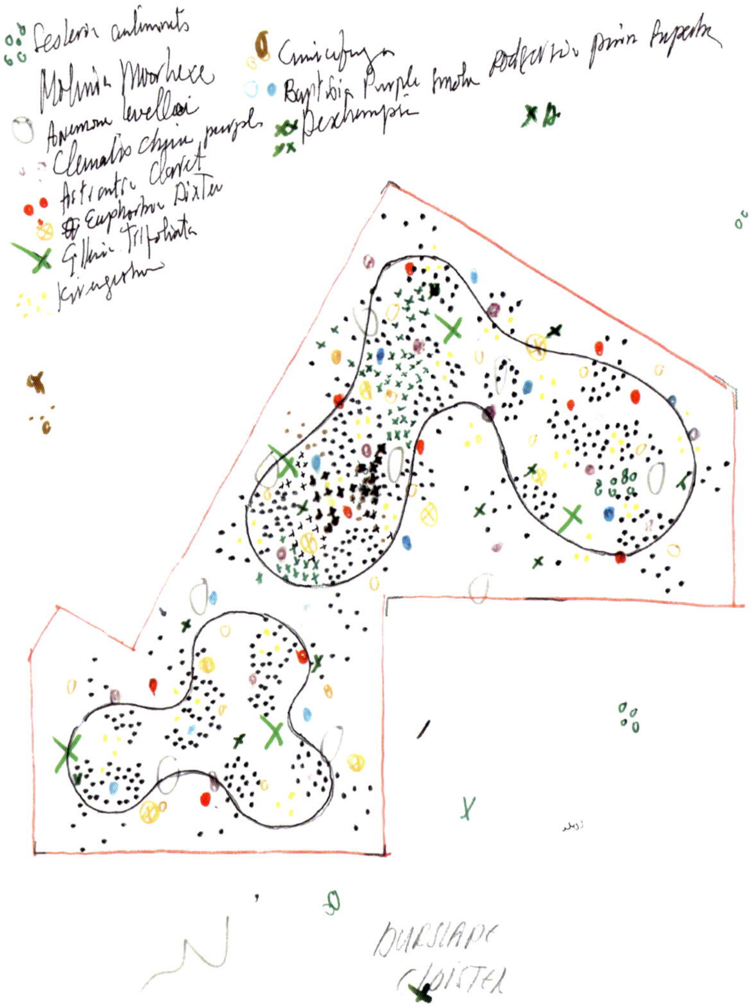

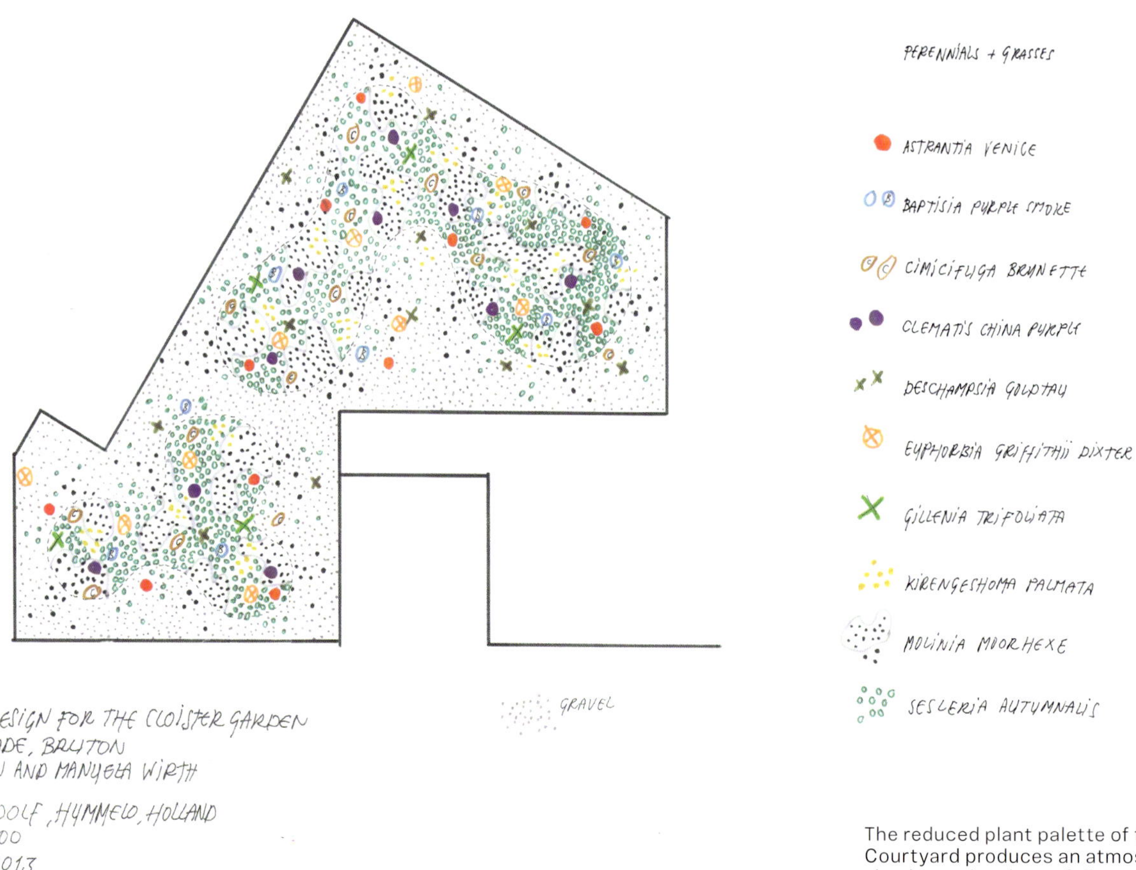

The reduced plant palette of the Cloister Courtyard produces an atmosphere that is simple, cool and very foliage-dominated, with grass *Sesleria autumnalis* playing a key role as a matrix plant.

Four seasonal views of Durslade Farm, from top to bottom, left to right: in spring, *Darmera peltata* flowering in the wet area, with *Caltha palustris* behind; in late summer, with plant height well established; in autumn, with the seedheads of *Phlomis russeliana* highlighted against the grass *Deschampsia cespitosa*; and in winter, hoar frost magically transforms perennial seedheads. *Overleaf*: The late autumn aspect of Oudolf Field. Seedheads and muted colours predominate; a good opportunity to appreciate the impact of repeated plant clumps.

De Vlinderhof

Máximapark, Utrecht, The Netherlands
2013–14

This planting in Máximapark, on the west side of Utrecht, is very much a public project. It was conceived when a local community figure was surprised to learn that there were so few Piet Oudolf projects that people could visit in his home country. A grant from the City of Utrecht funded the project, but strictly on the basis that it would be maintained by volunteers, which it is, and very successfully.

Máximapark's vast size and its combination of broad paths and relatively new areas of tree planting can be somewhat disorientating. Vlinderhof offers a safe haven, a place apart from the world, with its sheltering boundary of shrubs, some of which are integrated into the perennial planting. Most of the garden is in the sun, but one end is planted for shade, as young trees are expected gradually to change the conditions there. As with several other similar public projects, the paths here offer a multiplicity of ways to explore the garden without any one being obviously prioritized – one outcome of which is that plants and plant combinations may be encountered in several places or from several directions. The paths on both sides of the central bike lane encourage a slightly aimless wandering, relaxing and meditation. A rise in the ground at one end offers an opportunity for an overview.

A misty autumn day in Vlinderhof. The shrubs planted around the outside integrate the area into its largely wooded surroundings.

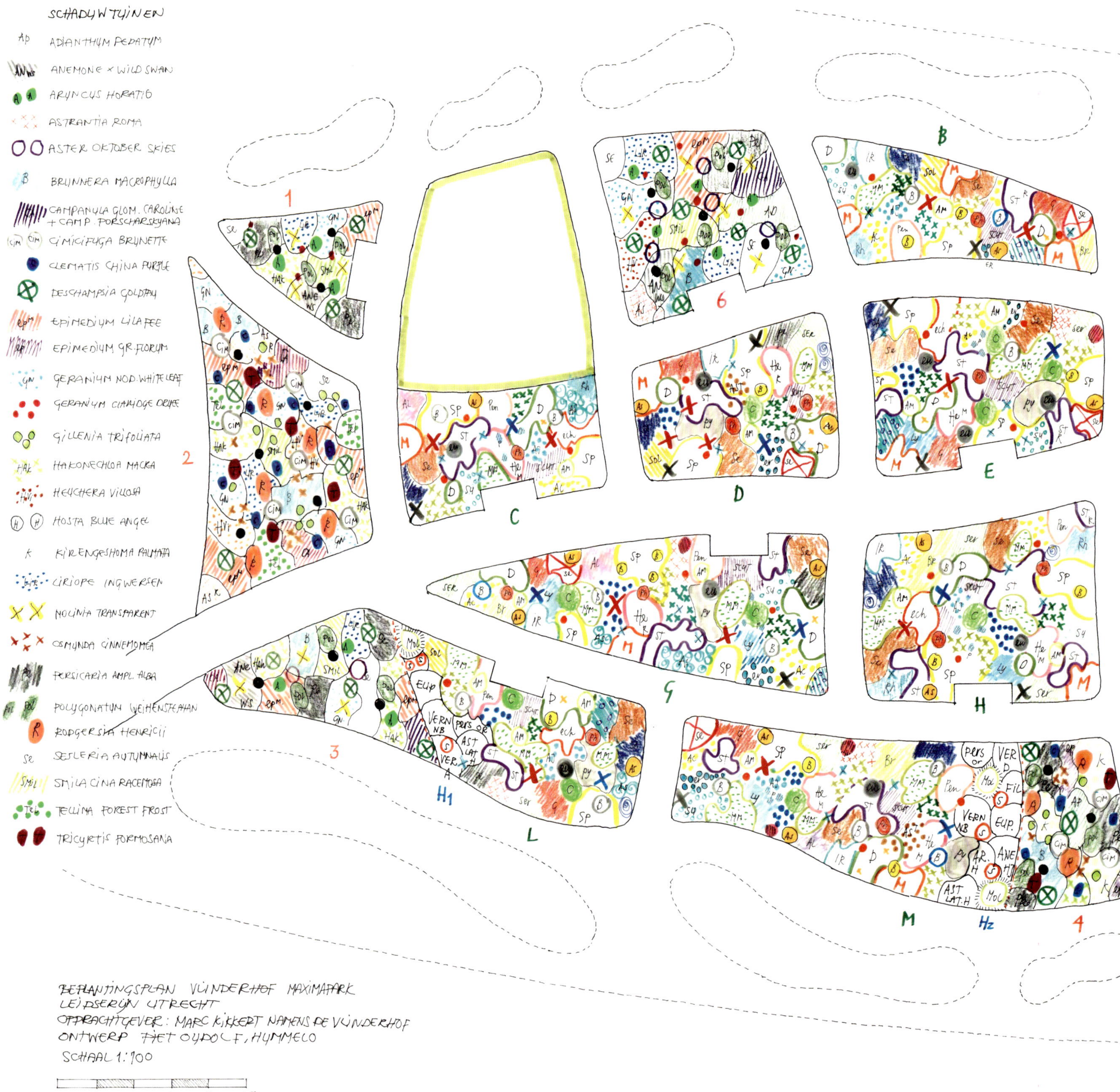

Planting plan for Vlinderhof. The more shaded areas are on the left and lower right (see beds 2, 4 and 5), and these plantings feature shrubs. There are almost a hundred species and varieties here, although they are far from evenly distributed, with most of the plant mass taken up by those with the longest season of good structure. Shorter-season plants in reduced quantities add splashes of interest.

VLINDERHOF
ACHILLEA MOONSHINE
ALLIUM SUMMER BEAUTY
AMSONIA HUBRICHTII
AMSONIA TAB.VAR. SALICIFOLIA
ASCLEPIAS TUBEROSA
ASTER LITTLE CARLOW
ASTER TWILIGHT
ASTER SPECTABILIS
ASTER TART JINDAY
ASTRANTIA CLARET
BAPTISIA CAROLINA MOONLIGHT
BAPTISIA PURPLE SMOKE
BAPTISIA ALBA
BRIZA MEDIA
CALAMAGROSTIS BRACHYTRICHA
DESCHAMPSIA GOLDTAU
ECHINACEA FATAL ATTRACTION
ECHINOPS VEITCH BLUE
ERYNGIUM YUCCIFOLIUM + CALAMINTHA NEP. SSP NEPETA
ERYNGIUM BOURGATII
EUPATORIUM MAC. SNOWBALL
EUPHORBIA FIRE GLOW
FESTUCA MAIREI
GERANIUM ROZANNE
GERANIUM REBECCA MOSS
GERANIUM PSILOSTEMON
HELENIUM MOERHEIM BEAUTY
HELENIUM RUBINZWERG
IRIS PERRY'S BLUE
LIATRIS SPICATA
LIMONIUM LATIFOLIUM + ORIGANUM HERRENHAUSEN
LYTHRUM VIRG. SWIRL
MOLINIA MOORHEXE + MELICA CILIATA
MONARDA BRADBURIANA
PANICUM VIRG. SHENANDOAH
PENSTEMON HUSKER'S RED
PEROVSKIA LITTLE SPIRE
PHLOMIS LITTLE AMAZONE
PYCNANTHEMUM MUTICUM + ECHINACEA PALLIDA
RHAZYA ORIENTALIS
SALVIA PURPLE RAIN
SALVIA PINK DELIGHT
SCUTELLARIA INCANA
SEDUM MATRONA
SEDUM RED CAULI
SELINUM WALLICHIANUM
SERRATULA SEOANEI
SPOROBOLUS HETEROLEPIS
STACHYS OFF. ROSEA
STACHYS OFF HUMMELO
SUCCISA PRATENSIS
PERSICARIA AMPL. ALBA
SOLIDAGO GOLDEN MOSA
HOGE BORDER
ANEMONE HONORINE JOBERT
ASTER LAT. HORIZONTALIS
EUPATORIUM RIESENSCHIRM
FILIPENDULA R. VEN. MAGNIFICA
MOLINIA TRANSPARENT
ARUNCUS HORATIO
PERSICARIA ORANGE FIELD
SANGUISORBA BLACKTHORN
VERNONIA NOVEBORACIENSIS
VERONICASTRUM ADORATION
VERONICASTRUM DIANE
EXTRA. KNAUTIA ARVENSIS, PAPAVER KARINE
BOMEN:
CERCIS CANADENSIS
CHIONANTHUS RETUSUS
GYMNOCLADUS DIOICUS
MAGNOLIA SOULANGEANA
PHELLODENDRON AMURENSE

Clockwise from upper left: *Eupatorium maculatum* 'Snowball' makes a great late season silhouette plant; the shaded end of the perennial garden, with *Tricyrtis formosana* 'Dark Form'; woody plants such as *Nyssa sylvatica* enable autumn colour to be an additional feature of a planting.

Private Garden
Berkshire, UK
2016–

A commission for a private garden in southern England is an example of the kind of long-term collaboration Piet has increasingly used in his later work. Here, Chris and Toby Marchant, who had previously run a specialist perennial nursery, have been intimately involved in the plant procurement and laying out of plants. Since Piet does not have a conventional design practice with an office and staff, his workload needs a different kind of model, one based on collaboration with other people who have experience of high-quality perennial design of their own or, as here, who have an intimate knowledge of the plants and a track record of the successful implementation of complex planting projects. Indeed, these collaborators usually enjoy considerable acclaim within the industry.

The garden is very obviously linked to its landscape of fields and mature oaks, though there is an absence of obvious boundaries. In many more traditional gardens, this creates a sense of disjuncture and a need to clearly state the separation. But with perennials and grasses, whose structures and textures are similar to those of wild plants in the hedgerows, it is much easier to segue into country landscapes. This is an ongoing project, with perennial planting tying together various buildings and garden features, such as a natural swimming pool and driveway.

Early winter is the best time to appreciate texture and structure in perennial gardens. Here there is a classic effect, the contrasting of dark, defined seedheads with pale, softly textured grasses.

Planting plan of this private garden, still a work in progress. This perennial planting surrounds the natural swimming pool and provides a link to other garden areas.

DUM MATRONA
SELI LIBANOTICUM
POROBOLUS HETEROLEPIS
ACHYS HUMMELO
LVIA PURPLE RAIN
STACHYS MON ROSEA
STIPA LESSINGIANA
SUCCISA PRATENSIS
INDIVIDUAL PLANTS:
ASTER OKTOBERSKIES
BAPTISIA PURPLE SMOKE
PAPAVER PERRY'S WHITE
SANGUISORBA MENZIESII
SELINUM WALLICHIANUM
AMSONIA V. SALICIFOLIA
C
SCALE 1:100 METRIC
0
2
4
6
8
10 M

Clockwise from top left: Autumn colour of *Amsonia hubrichtii* in the perennial meadow planting; the strong 'negative space' impact of *Sedum* 'Matrona'; flower spikes of *Actaea* 'Queen of Sheba'.

Meadow Garden at Delaware Botanic Gardens
Pepper Creek, Dagsboro, Delaware, USA
2017–19

Having opened in 2019, this relatively new botanic garden is unusual in that it is run and funded by volunteers, having started with a community group some years previously. In an area with few public gardens, this volunteer-run organization masterminded the creation of the Delaware Botanic Gardens, which compromises a folly garden and woodland gardens among others, with the aim of becoming a landmark public garden for the region and a tourist destination for Delaware. Piet was invited to design a Meadow Garden in 2015, which was planted in three phases, starting in 2017. The 0.8 hectare (2 acre) open space is very much an ornamental gardened centrepiece for an institution that focuses on regional ecology, with much of its space being natural or semi-natural habitat. The planting is unusual in that a high proportion of it uses the matrix style, the ground mass being three locally native grass species.

The planting includes a very high number of native species, higher than usual for Piet's plantings, but appropriate for an institution heavily focused on local flora. The non-native species used are very much there for a purpose, for example, some *Geranium* cultivars that produce flowers earlier than many of the natives used, and two cultivars of *Kniphofia*, a South African genus he only rarely uses, but with dramatic flower spikes distinctly different to any North American natives. The planting of the garden was carried out by around a hundred collaborators and volunteers who have worked with Piet before, playing a crucial role in marshalling plants and people. While most of the volunteers were local, some came a long way, even from Europe, to share their horticultural and organizational skills.

Seedheads of *Echinacea pallida* are silhouetted against *Amsonia hubrichtii* in the first stages of its dramatic autumn colour change. *Overleaf*: The Meadow Garden: a matrix planting plan above, and block planting plan below, although with two small areas of matrix. Right, the garden in early autumn.

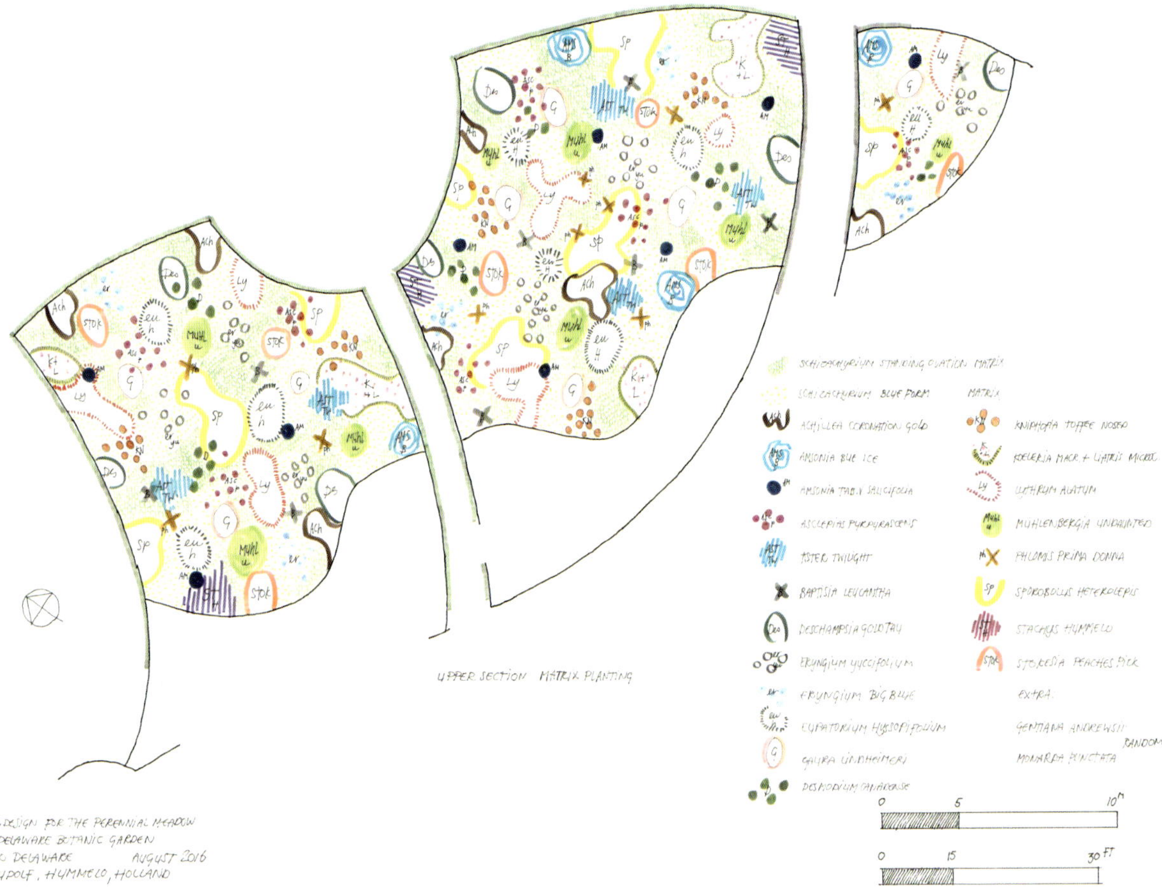
UPPER SECTION MATRIX PLANTING
PLANTING DESIGN FOR THE PERENNIAL MEADOW
AT THE DELAWARE BOTANIC GARDEN
DAGSBORO, DELAWARE AUGUST 2016
PIET OUDOLF, HUMMELO, HOLLAND

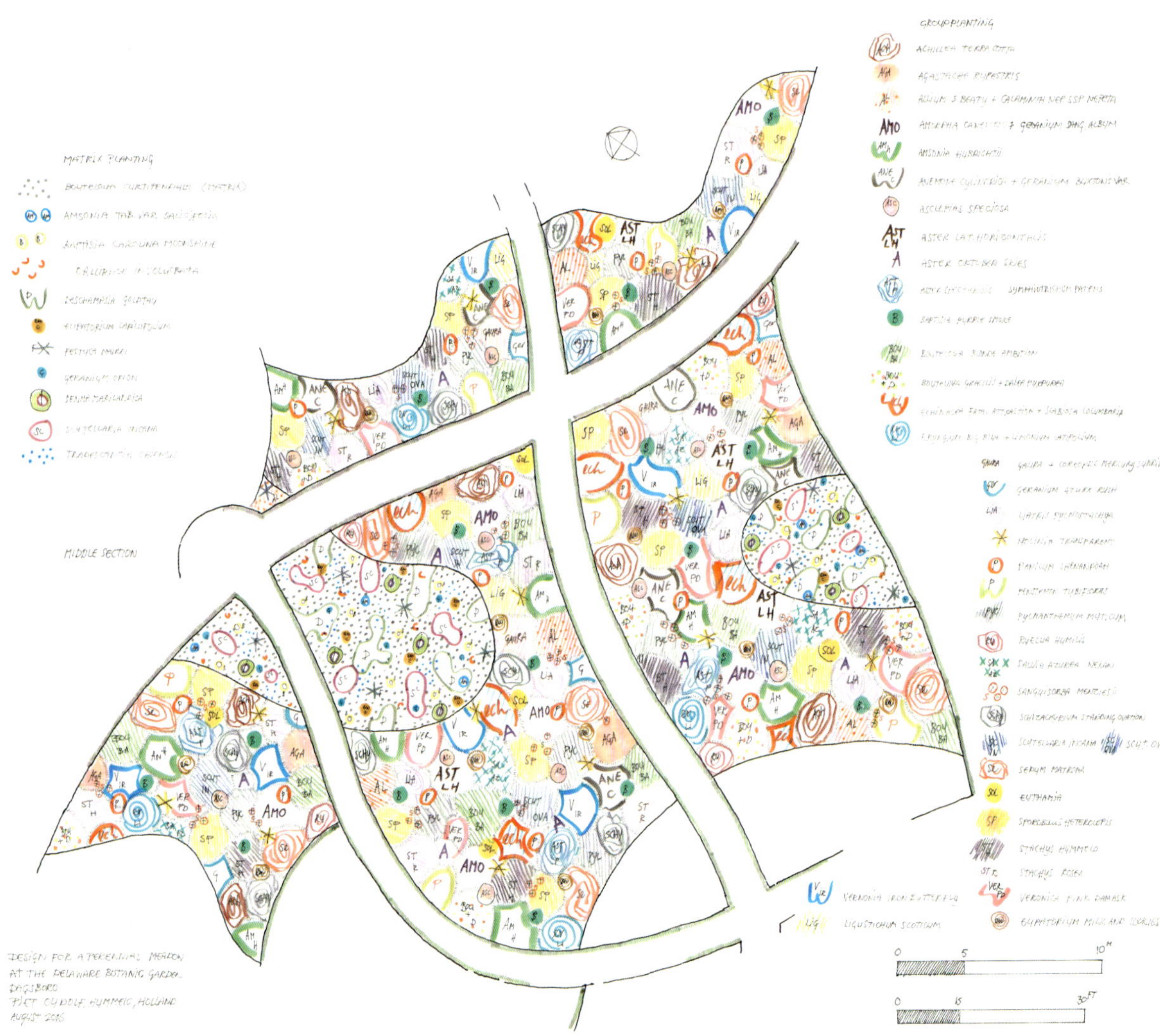
MATRIX PLANTING
GROUP PLANTING
MIDDLE SECTION
DESIGN FOR A PERENNIAL MEADOW
AT THE DELAWARE BOTANIC GARDEN
DAGSBORO
PIET OUDOLF HUMMELO, HOLLAND
AUGUST 2016

Singer Laren

Laren, The Netherlands
2018

Singer Laren was founded in 1956 to present and preserve the art collection of the American artist William Henry Singer (1868–1943) and his wife, Anna (1873–1962); it underwent a major reconstruction in 2017, when Piet was invited to design new gardens both to connect various parts of the museum and the theatre, and to create spaces where contemporary sculpture could be exhibited. It might be argued that the soft textures of many of the plants he uses, and the lack of emphatic form or strong colour, makes his style eminently suitable for complementing sculpture. Needless to say, plant height was a consideration, in order not to obscure artworks.

The courtyard of the museum forms the main part of the garden, divided into twelve central beds of slightly varying size and shape, with a number of others around the edge. Block and scatter planting is used throughout, except for the central bed, which forms a small matrix planting using *Sesleria autumnalis*, probably the most successful of the various species Piet, and others, have tried for this planting style. It is valuable for a growth rate fast enough to give results but not so fast as to overwhelm other plants. Piet used over a hundred different plants in the garden, with groups of different species in each section. 'It's a traditional way of planting,' he says, 'but the plant species used and the combinations mean it won't look traditional.' A roof garden connects the main garden to a newly built theatre building; accessible from the foyer, the roof garden uses a different plant palette, one better adapted to cope with the stress of shallow soil; it does however need irrigation if it is to perform well.

The Sculpture Garden in early summer. The *Sesleria* matrix, which forms the central bed, is to the left. All the planting here is relatively low, in order to provide a suitable backdrop and foundation for the artworks and sculpture.

Sesleria

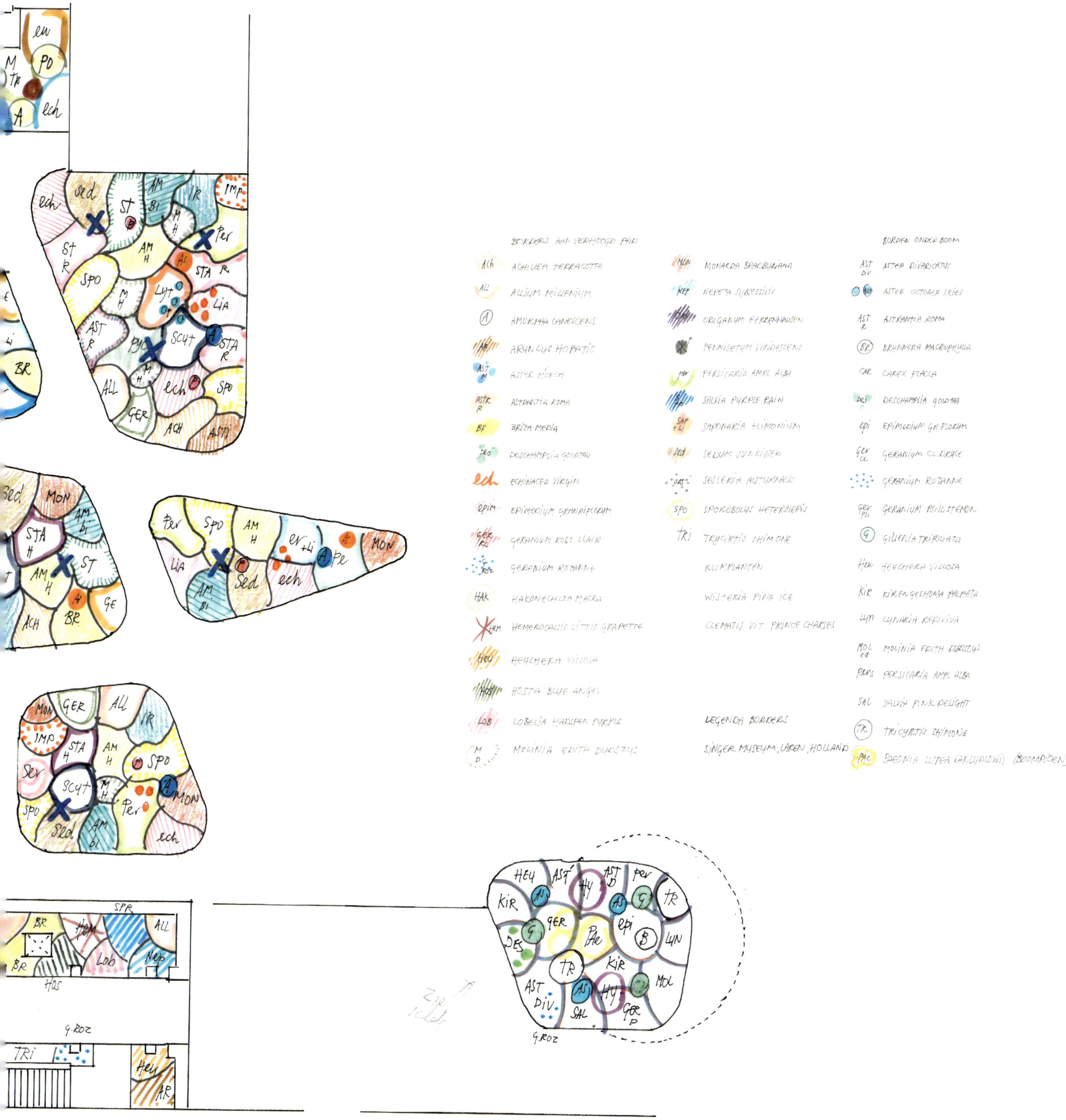

Planting plan for the main Sculpture Garden; the central bed using a matrix planting of *Sesleria autumnalis* is clearly visible, while block and scatter-style planting is used for the surrounding beds.

Views of the Sculpture Garden (above) and Roof Garden (top, with the planting plan and legend for the roof garden opposite). *Stachys officinalis* and grass *Stipa tenuissima* are among the lower-growing plant selection here.

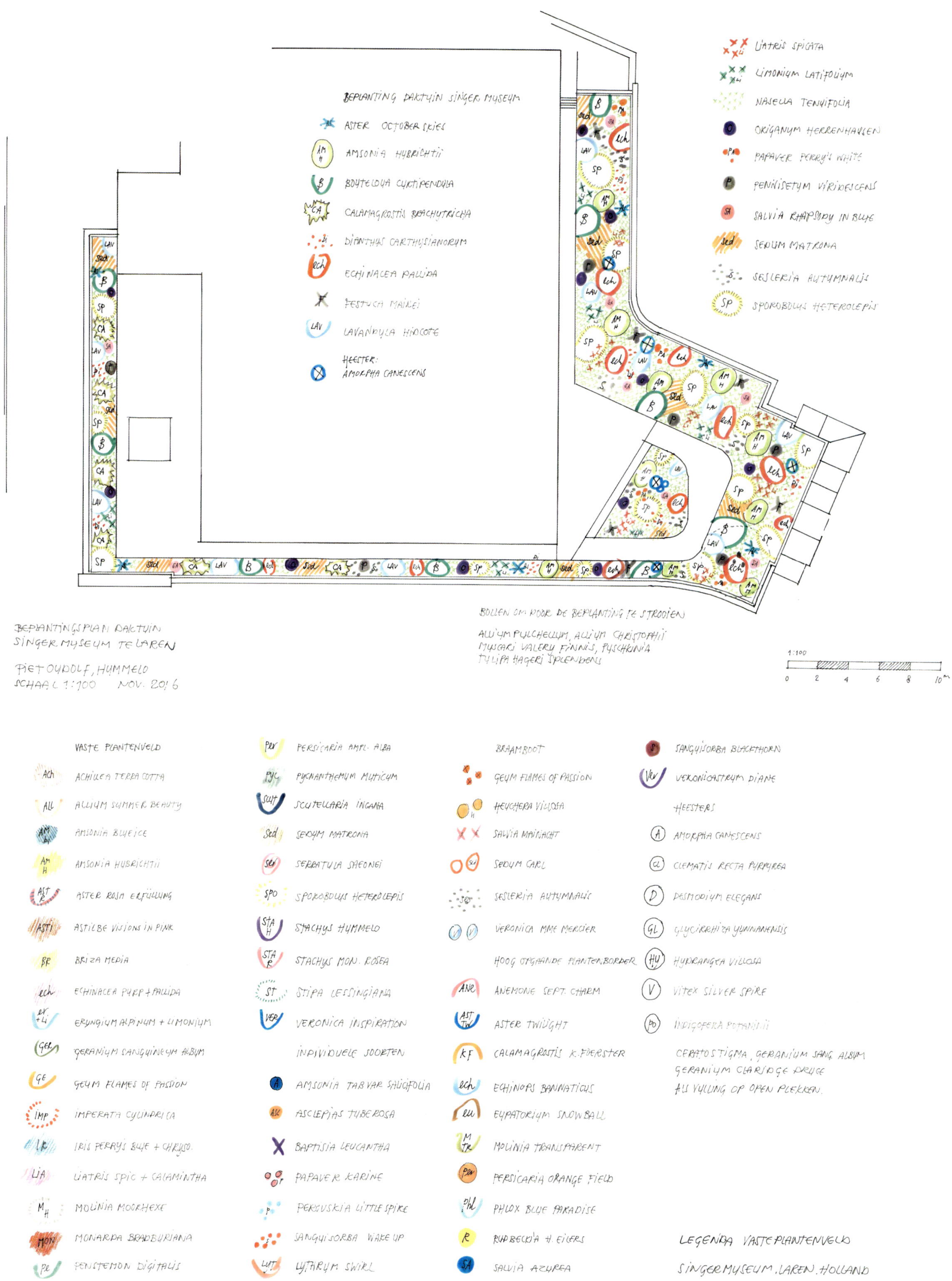
BEPLANTING DAKTUIN SINGER MUSEUM
ASTER OCTOBER SKIES
AMSONIA HUBRICHTII
BOUTELOUA CURTIPENDULA
CALAMAGROSTIS BRACHYTRICHA
DIANTHUS CARTHUSIANORUM
ECHINACEA PALLIDA
FESTUCA MAIREI
LAVANDULA HIDCOTE
HEESTER:
AMORPHA CANESCENS
LIATRIS SPICATA
LIMONIUM LATIFOLIUM
NASELLA TENUIFOLIA
ORIGANUM HERRENHAUSEN
PAPAVER PERRY'S WHITE
PENNISETUM VIRIDESCENS
SALVIA RHAPSODY IN BLUE
SEDUM MATRONA
SESLERIA AUTUMNALIS
SPOROBOLUS HETEROLEPIS
BEPLANTINGSPLAN DAKTUIN
SINGER MUSEUM TE LAREN
PIET OUDOLF, HUMMELO
SCHAAL 1:100 NOV. 2016
BOLLEN OM DOOR DE BEPLANTING TE STROOIEN
ALLIUM PULCHELLUM, ALLIUM CHRISTOPHII
MUSCARI VALERY FINNIS, PUSCHKINIA
TULIPA HAGERI SPLENDENS
1:100
0 2 4 6 8 10m
VASTE PLANTENVELD
ACHILLEA TERRA COTTA
ALLIUM SUMMER BEAUTY
AMSONIA BLUE ICE
AMSONIA HUBRICHTII
ASTER ROSA ERFÜLLUNG
ASTILBE VISIONS IN PINK
BRIZA MEDIA
ECHINACEA PURP + PALLIDA
ERYNGIUM ALPINUM + LIMONIUM
GERANIUM SANGUINEUM ALBUM
GEUM FLAMES OF PASSION
IMPERATA CYLINDRICA
IRIS PERRYS BLUE + CHRYSO.
LIATRIS SPIC + CALAMINTHA
MOLINIA MOORHEXE
MONARDA BRADBURIANA
PENSTEMON DIGITALIS
PERSICARIA AMPL. ALBA
PYCNANTHEMUM MUTICUM
SCUTELLARIA INCANA
SEDUM MATRONA
SERRATULA SHAWII
SPOROBOLUS HETEROLEPIS
STACHYS HUMMELO
STACHYS MON. ROSEA
STIPA LESSINGIANA
VERONICA INSPIRATION
INDIVIDUELE SOORTEN
AMSONIA TAB VAR. SALICIFOLIA
ASCLEPIAS TUBEROSA
BAPTISIA LEUCANTHA
PAPAVER KARINE
PEROVSKIA LITTLE SPIRE
SANGUISORBA WAKE UP
LYTHRUM SWIRL
BRAMBOOT
GEUM FLAMES OF PASSION
HEUCHERA VILLOSA
SALVIA MAINACHT
SEDUM CARL
SESLERIA AUTUMNALIS
VERONICA MME MERCIER
HOOG OPGAANDE PLANTENBORDER
ANEMONE SEPT. CHARM
ASTER TWILIGHT
CALAMAGROSTIS K. FOERSTER
ECHINOPS BANNATICUS
EUPATORIUM SNOWBALL
MOLINIA TRANSPARENT
PERSICARIA ORANGE FIELD
PHLOX BLUE PARADISE
RUDBECKIA H. EILERS
SALVIA AZUREA
SANGUISORBA BLACKTHORN
VERONICASTRUM DIANE
HEESTERS
AMORPHA CANESCENS
CLEMATIS RECTA PURPUREA
DESMODIUM ELEGANS
GLYCYRRHIZA YUNNANENSIS
HYDRANGEA VILLOSA
VITEX SILVER SPIRE
INDIGOFERA POTANINII
CERATOSTIGMA, GERANIUM SANG. ALBUM
GERANIUM CLARIDGE DRUCE
ALS VULLING OP OPEN PLEKKEN
LEGENDA VASTE PLANTENVELD
SINGER MUSEUM, LAREN, HOLLAND

Noma
Copenhagen, Denmark
2019

Founded by Chef René Redzepi, Noma is one of the world's most innovative restaurants, known for its New Nordic cuisine. It is situated on a strip of land overlooking one of the many bodies of water that break up the city of Copenhagen. A planting by Piet lines each side of a long, straight path, with the buildings and greenhouses of the restaurant along one side and a backing of reeds and the water on the other. Many diners choose to walk this garden path before or after their meal. At one point, in an area between the restaurant and the test kitchens, the planting takes on a more experimental character. Every year a discussion takes place between Piet and the Noma team about how this area might be adapted. At the end of the path is a small lawn and a strip of woodland, with a shaded area, known as the Fern Garden, to one side – cool green foliage and ground cover predominate.

As with many of his projects, this is a garden in which Piet takes a particular personal interest, and he and the head gardener talk frequently. The annual changes necessitate an ongoing discussion between the two, but also allow the garden to evolve and develop in keeping with Piet's original vision.

Symphyotrichum 'Little Carlow' (on the left) flowers from late summer in the garden and provides autumn colour into November before the wintery, brown and gold hues and various seedheads dominate. The restaurant building, seen here on the left, and the sunflowers in the middle distance either side of the path mark the area reserved for annual and experimental plantings.

An early sketch for the entrance (left), dominated by a *Sesleria autumnalis* matrix, and (right) a draft plan for the shaded area at the rear of the restaurant, with woodland shrubs such as species of *Hamamelis*, ferns and other forest-floor perennials.

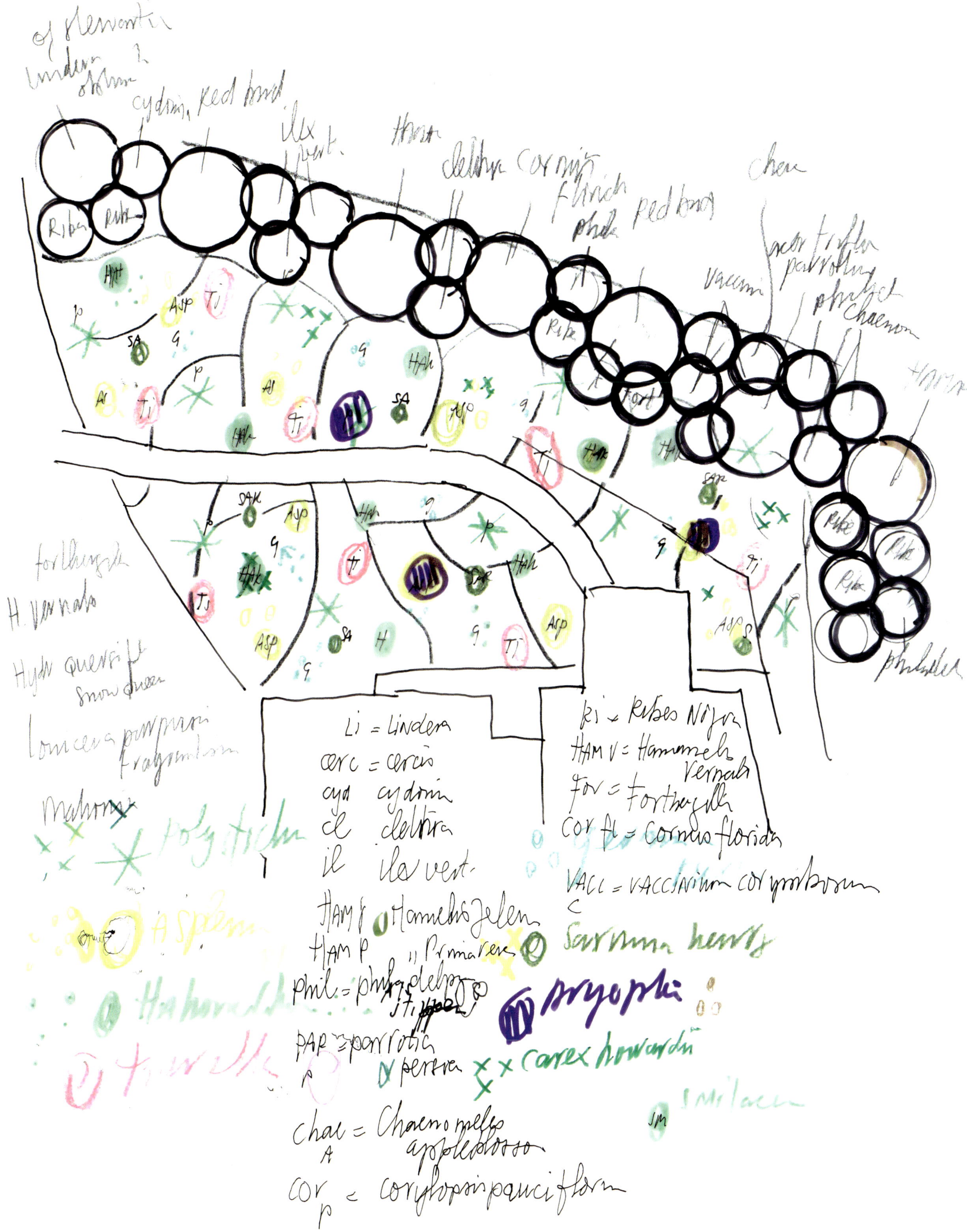

Sketch for the long garden (right) and final planting plan (below). The blank middle area is reserved for annuals.

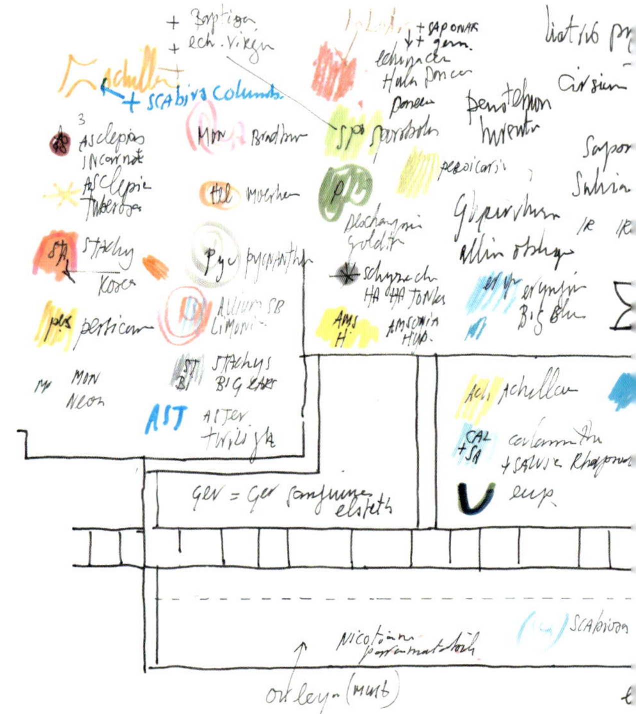

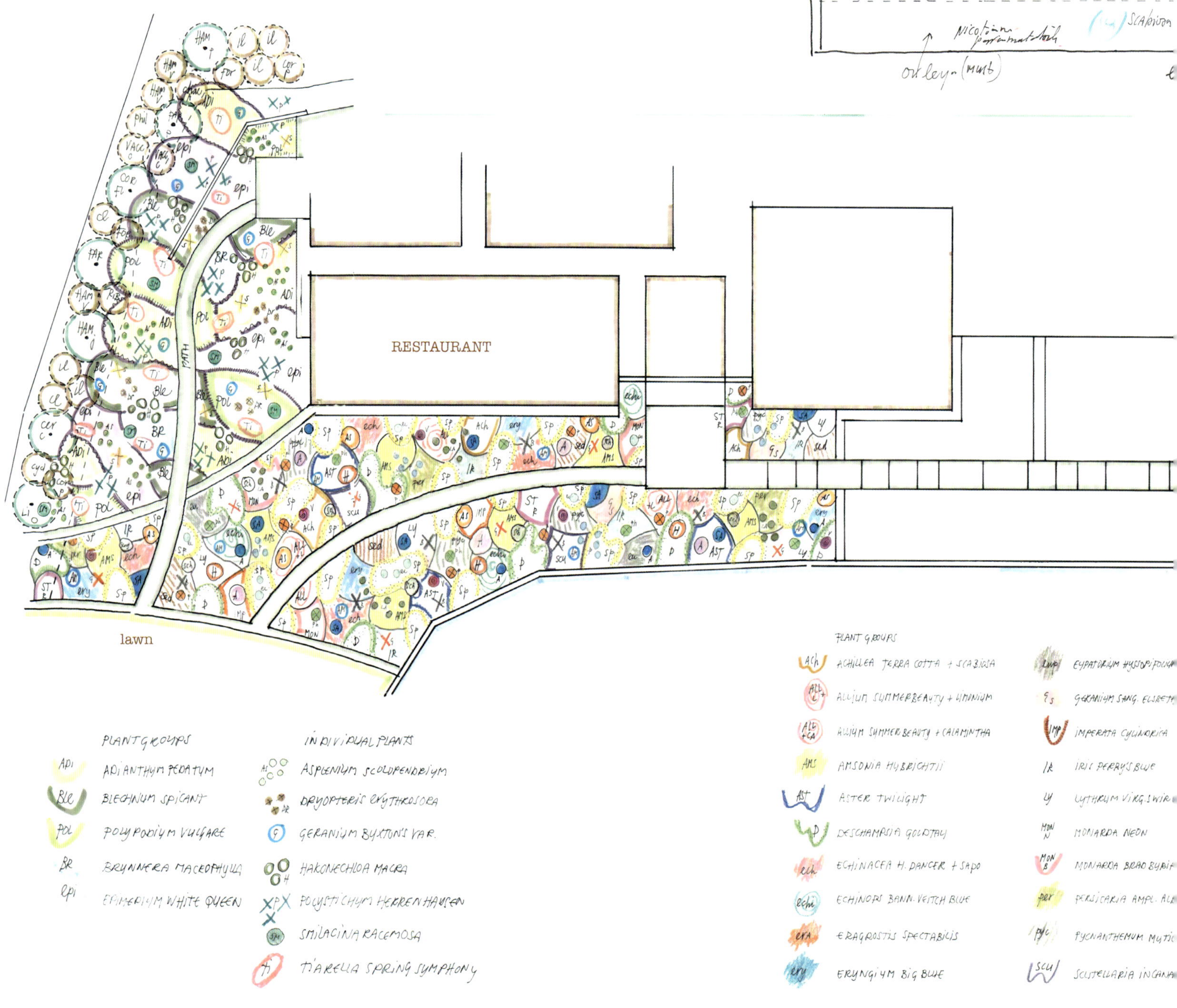

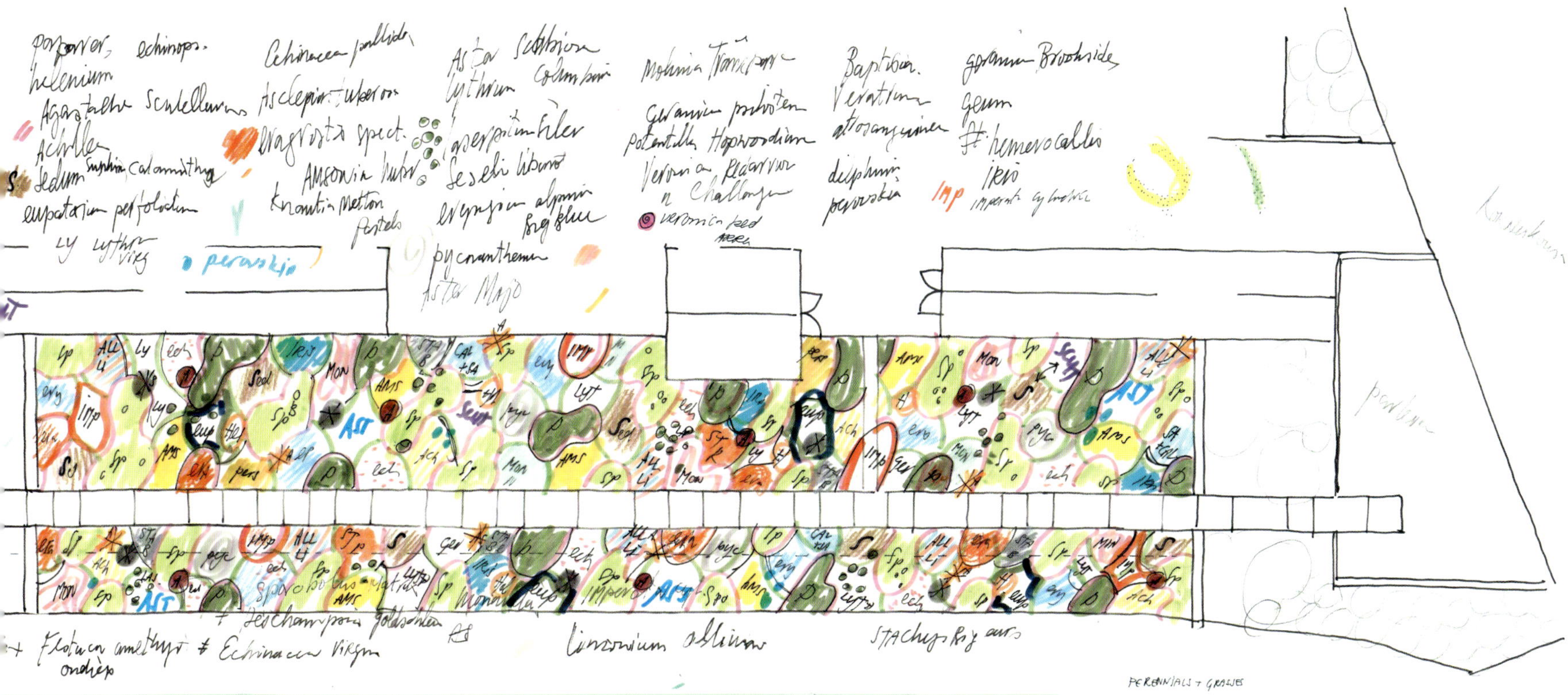

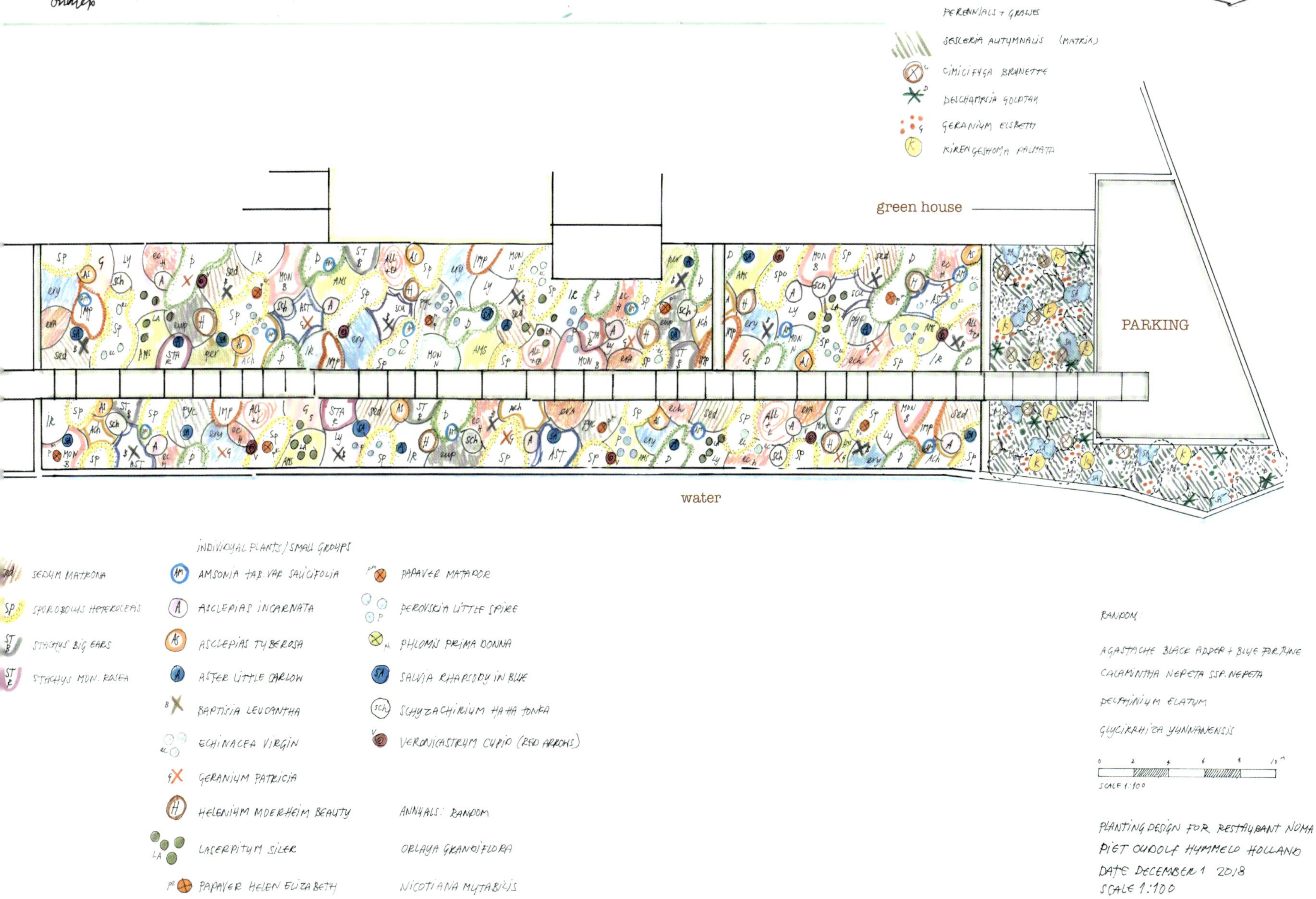

PERENNIALS + GRASSES
SESLERIA AUTUMNALIS (MATRIX)
CIMICIFUGA BRUNETTE
DESCHAMPSIA GOLDTAU
GERANIUM ELSBETH
KIRENGESHOMA PALMATA
green house
PARKING
water
INDIVIDUAL PLANTS / SMALL GROUPS
SEDUM MATRONA
SPOROBOLUS HETEROLEPIS
STACHYS BIG EARS
STACHYS MON. ROSEA
AMSONIA TAB. VAR SALICIFOLIA
ASCLEPIAS INCARNATA
ASCLEPIAS TUBEROSA
ASTER LITTLE CARLOW
BAPTISIA LEUCANTHA
ECHINACEA VIRGIN
GERANIUM PATRICIA
HELENIUM MOERHEIM BEAUTY
LASERPITIUM SILER
PAPAVER HELEN ELIZABETH
PAPAVER MATADOR
PEROVSKIA LITTLE SPIRE
PHLOMIS PRIMA DONNA
SALVIA RHAPSODY IN BLUE
SCHYZACHIRIUM HA HA TONKA
VERONICASTRUM CUPID (RED ARROWS)
ANNUALS: RANDOM
ORLAYA GRANDIFLORA
NICOTIANA MUTABILIS
RANDOM
AGASTACHE BLACK ADDER + BLUE FORTUNE
CALAMINTHA NEPETA SSP. NEPETA
DELPHINIUM ELATUM
GLYCIRRHIZA YUNNANENSIS
SCALE 1:100
PLANTING DESIGN FOR RESTAURANT NOMA
PIET OUDOLF HUMMELO HOLLAND
DATE DECEMBER 1 2018
SCALE 1:100

The annual area (top left) also features edibles with the perennials; a range of *Echinops* and *Echinacea* cultivars are used in the planting by the glasshouse (top right); and by autumn, seasonal grasses take over from flowering perennials and provide interest, especially red *Imperata cylindrica* (lower image). In winter, seedheads are prominent (opposite).

Maggie's at the Royal Marsden Hospital
London, UK
2019

Set in the suburbs of London, this garden for a cancer care charity surrounds and complements a building designed by Ab Rogers Design. The overall role of the garden is to provide a place of respite and healing for patients, their families and friends; part of the mission of the Maggie's Centres is to combine innovative architecture with gardens created by leading designers.

The garden is divided into four interconnected zones, the basic character of which is determined by how much sunlight they get. The narrow Dynamic Garden makes the most of the sun-loving perennials that are at the heart of Piet's design style, with a dramatic growth from bulbs to waist-height perennials and grasses by the end of the season. The more extensive Shaded Garden makes use of small trees, shrubs, ground-covering perennials, ferns and a few grasses that have some shade tolerance. The Woodland Garden, divided into two areas, uses existing trees with new shrub planting to create privacy and a calm, contemplative atmosphere, while another area acts as screening. Around the entrance to the building there is a fourth area, a narrow border with ten species of summer-flowering perennials, although all are also noted for their strong presence in winter; they include *Glycyrrhiza yunnanensis*, a species of liquorice, always popular for its spiky seedheads in winter, carried on a bushy plant. Tight spaces like this offer an opportunity to see what designers can do in small gardens, one of the frustrations of many home gardeners being that the more successful designers become, the larger their projects tend to be, and therefore the less relevant to the interests of the majority of the public who follow their work.

A woodland planting creates a cool-themed area in the garden at Maggie's. The presence of existing trees guided planting at the entrance towards this woodland feel; the large leaves are *Salvia glutinosa*, the young shrubs a variety of *Amelanchier*. Other species include ferns and a wide selection of woodland-edge grasses and sedges.

Planting plan around the building. Light shade offers an opportunity to bring together plants of both open and woodland conditions. The fern *Polystichum setiferum* 'Herrenhausen' is scattered as a rhythmic element, for which it is ideal, being more or less evergreen and with a tolerance for different light levels. Other plants scattered among the block planting include an early-flowering poppy (*Papaver orientale* 'Juliane') and late-performing *Aster* 'October Skies'.

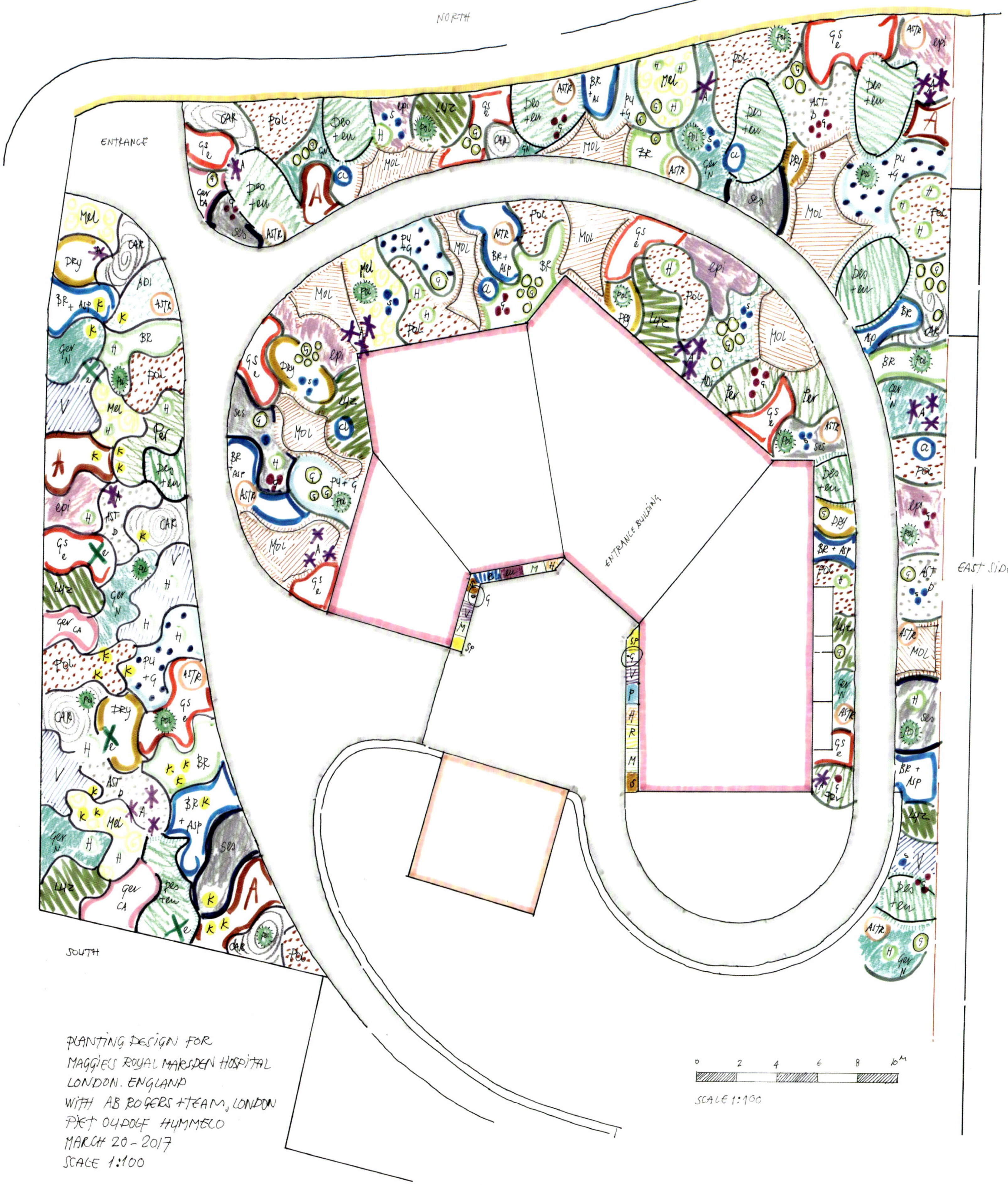
NORTH
ENTRANCE
ENTRANCE BUILDING
EAST SIDE
SOUTH
PLANTING DESIGN FOR
MAGGIES ROYAL MARSDEN HOSPITAL
LONDON. ENGLAND
WITH AB ROGERS + TEAM, LONDON
PIET OUDOLF HUMMELO
MARCH 20 - 2017
SCALE 1:100
0 2 4 6 8 10M
SCALE 1:100

Clockwise from upper left: Spring bulbs, mostly species tulips, are used throughout, conveniently dying back by the time the foliage of the perennials covers the ground; autumnal seedheads of *Echinacea pallida*; the fresh green of *Sesleria autumnalis*, a favourite grass for matrix planting, with seedheads of native grass *Deschampsia cespitosa* behind; a plan of the Woodland Garden, with small tree and shrub planting acting as screening and, to the left, an area of *Sporobolus heterolepis* matrix. The implementation of this project was carried out by Chris and Toby Marchant, former owners of a well-known wholesale perennial nursery. Piet went on to use them to implement other projects (see page 157).

In Conversation: Piet Oudolf, Hans Ulrich Obrist and Tino Sehgal

HUO: Good morning, Piet. I have a lot of questions, but I suppose the first is, how did you come to gardens and the design of landscape – or how did gardens come to you? Was there a sudden revelation or was it a gradual process?

PO: No, it was not a gradual process. It happened suddenly after my twenty-fifth birthday. I spent one Christmas in a garden centre, selling Christmas trees, and they asked me to work in the nursery. It occurred to me that this was something I really liked. I went to evening classes for a few years, and in the meantime I worked for garden construction companies and for gardeners. When I finished the evening classes, in 1975, I started a small garden design company. My passion for plants grew very fast. I was starting to travel to England, to gardens and nurseries, and I collected every plant that I liked. I had no garden myself, so I planted them in my mother's garden at the restaurant; I created a garden there for the first time.

HUO: Who are the landscape architects or garden pioneers who inspired you?

PO: It was English gardens that inspired me at the beginning. In the late 1970s, Anja and I started to travel to England. I was obsessed by plants, by English gardens and parks, by the decorative value of plants. Nature wasn't part of my thinking. It was more that I wanted to see as many plants in as many places as possible. I grew up in a nature reserve, but I never realized that that might have been an influence until I was older.

At the beginning of my career, plants spoke to me as characters, and I wanted to do something with them. I thought in terms of beauty, not art – how you can put plants together as personalities to create a beautiful picture. I felt I could do work that showed something of myself to people.

A big change was when we moved from the city of Haarlem to the countryside where we live now. Because I felt such a strong need to grow plants, we bought a derelict farmhouse and opened a nursery and trial gardens. The trial gardens gave me the sort of experience that a lot of landscape designers never get: working with plants on your own site, trying them out to see how they perform, what character they have. During that period we met people involved in the native plant movement, and that made me decide to get away from the dogmatic style of gardening I had followed up to then.

English gardening is a lot about decoration, and about doing the right thing at the right time. There were calendars and books that taught you what to do when, and I wanted to free myself from that. The people I met, like Henk Gerritsen and Rob Leopold, were strongly into using wild, native plants in their gardens, and they were interested in what we were growing in our nursery. We had plants that they probably knew only from seeing them in nature when they were travelling in Europe or Asia. Our nursery became famous, a hub for plant lovers.

HUO: You've discovered many new plant varieties in your long career, which must be incredibly exciting.

PO: Discoveries mean selection, finding new plants in a world where all the main nurseries just grow what sells. We discovered plants that stood out, were more beautiful, had a better character, a nicer colour. You need an eye for a good plant. You see a range of seedlings and you say, OK, that is one I can use.

HUO: Apart from your knowledge of plants, another thing that is very specific and special to you is your amazing drawings. Is that something you've always done?

PO: Not in the beginning, when the gardens we were designing were fairly small. Then bigger gardens came, and public gardens. That was when I had to put something on paper to show where particular plants would stand in combination with others. I started to make my garden designs on paper in the late 80s.

HUO: Piet, it would be great to hear more about your own garden at Hummelo, because I know you've left it to change freely with the seasons, in a very organic way. I've always been interested in the idea of how to negotiate control and not control, or organization and self-organization. Was that a methodology that was honed in this laboratory of your own garden?

PO: I've always seen gardening as control. You cannot make a garden without it. Traditional gardening is a sort of super-control of everything that happens there. We started to discuss what we could do to make gardens more attractive not only for ourselves but also as living things. We found that the best way was to leave plants to go to seed. We stopped cutting back the plants too early – the skeletons were attractive enough to leave. At that time we didn't even think about the birds, but only about aesthetics and how we could extend this beauty through the seasons. It is not that you lose control, but that you start to look differently. For someone who is not a gardener, it looks uncontrolled, but it's not – it's just that you look more than you act, rather than acting without thinking.

When I started to do public gardens like the High Line or the Lurie Garden at Millennium Park [pp.83 and 75], we had to work with gardeners and with plants that were, both of them, growing and developing. We had to see what would happen, and I think that is losing control without losing control, because we take control when necessary. Otherwise, it becomes wild. You're always looking for the edge of control. If plants are dying, we ask the gardeners what should be done. Maybe this plant doesn't work any more, because the trees have grown too big. It becomes a dialogue with the gardener or owner about how we can make the garden work in the future, and keep its beauty.

TS: So it's an aesthetics of a controlled letting go or an aesthetics of controlled non-control. I was in one of your gardens in Hamm in November, and I was struck by all these plants that were decaying and yet still sculptural. Of course it's planned, but on the other hand it's a plan to let go of control.

PO: Yes. When I see plants decaying in a setting like that, I'm still struck. A plant in a particular context can influence your idea of aesthetics. It's like a good painting: you can see it a hundred times, and it still touches you. But plants change, and if you don't control a planting, you won't like it any more. In ten years' time you'll say, they should have done something about this.

HUO: In that context, the 1990s was an important time for you in terms of invention. Could you tell us about where you first applied your inventions outside the garden at Hummelo?

PO: My first international garden was a commission in Sweden [see pp.247 and 249] in 1996. I met the director of parks at a conference, and he asked me if I could do a small garden near a library in Enköping, 35 minutes' drive from Stockholm. It was my first experience outside the Netherlands and outside my gardens. It was received very well, but really it was making the garden I had at home in a different place. I gained a lot of experience from it, because a few plants I used died; they were not able to survive the climate, were pushed out by other plants, or were short-lived. I moved from thinking about the characters of plants to thinking about how they would live together. How would they take care of each other? Are they short-lived or aggressive? Do they need more or less water? I had to learn a lot about drought or wet, shade or sun, and habitat.

HUO: Your work is often described as part of the New Perennial movement. Please tell us more about this movement.

PO: It was called the New Perennial movement, and in some places the Dutch Wave, because so many people came together in the 1980s and early 90s with the idea that we should do something different in gardens and public spaces. It's not that I don't use trees and shrubs, because in every garden of mine there are trees and shrubs, but the focus at that time was on perennials. It became a movement very slowly, and they call it one now because it was copied so much.

There have been a lot of books about naturalistic planting, especially since around 2010, but I think it all started in the 1980s. A group of people from Holland, Germany and England held symposia about perennial plants for public spaces and their usefulness in making gardens more attractive. In the meantime we started to see decay as part of gardening, because we began to see that everything that happens in the garden is in the context of the season. You don't expect flowers in the winter, and you don't expect decay in the spring. That is part of the thinking of this movement, that you don't vacuum your garden every day because you think it has to be clean. You look at your garden, and you think, OK, that looks good, although it is falling over. We were just following our own ideas about how a garden should look. We learned to garden with our eyes and senses.

HUO: When we collaborated on the garden for Peter Zumthor's Serpentine Pavilion in London, you explained to me the 70 per cent rule. There are two kinds of perennial: structural perennials and filler perennials. You said that 70 per cent should be structural and the rest filler.

PO: Well, it's not really a rule. It is more to say that not everything has to show off. There are plants that come into flower, and then go dormant in the summer, like some of the poppies. We call these filler plants, which fill the bottom storey a little, flower, then disappear. You need them to make the garden complete.

HUO: Then there is repetition. The plants you repeat often are *Salvia pratensis* varieties and *Aster oblongifolius* [*Symphyotrichum oblongifolium*] 'October Skies'.

PO: Yes, and many more. In public gardens there are only so many perennials that you can use. They have to be hardy, not aggressive, and long-lived. So particularly in bigger spaces, repetition is key: you come back to the same plants, maybe in different-sized groups. A plant such as the aster you mention flowers very late. If you repeat it, you see it scattered over the whole area. The same with the *Salvia*, which flowers very early. If you put something in the middle and it's only there, the eye will be pushed in that one direction, and I want the eye to flow through the garden.

TS: This reminds me also of the twentieth-century philosopher Hans-Georg Gadamer. Earlier you used the word 'decoration' in a negative sense, but there's also a positive side to decoration, as something embedded in its environment. Gadamer says that what decoration does – and what sculpture mostly doesn't do – is attract the eye but then pass it on. You just said planting should flow. That might be why I enjoy your gardens. There's one in particular – the Lurie Garden in Chicago – where you have this long strip of velvety purple, like a river of flowers. It's very relaxing and beautiful, so maybe this is what you mean by fluid.

PO: I still love English gardens, and I don't see decoration as negative. I still love this over-the-top thing.

HUO: Could you describe the process you go through when you start a project? How do you begin to landscape space, and what do you take into consideration in terms of surrounding architecture?

PO: Context. The context of a good garden is very important. Working in a city is completely different from working in the countryside. You can imagine that in a city like Chicago, with skyscrapers in the background, the context is so powerful that you have to counterbalance it with your planting. The planting should have an impact; people should feel it has something to say. On the other hand, my work at Hauser & Wirth Somerset is rural, and it feels completely different.

For the Lurie Garden in Chicago, we were asked to take part in a competition in 2000. I saw the list of people entering, and there were at least three that I thought I could not win against. The organizers suggested that I collaborate with one of the other entrants, and that's how I joined up with Kathryn Gustafson.

TS: What is your relation to landscape architecture, given that you really see yourself as a plant person, a gardener?

PO: Well, I did the masterplan and planting for the Vitra garden in Weil am Rhein [p.203], in Germany, in 2018–20, and I'm very happy with how I laid that out; and there's also Hauser & Wirth Somerset [p.131]. I'm happy with what I did, but I'm not a landscape architect by training. Also, landscape architects mostly have offices with many people, and I work on my own. Sometimes a project is too big and too complex for me to do everything myself, and that's why I joined with James Corner Field Operations for the High Line – also because I loved the project, and the intent of the people who wanted to rescue the site. It was one of the few times I joined a competition. So I guess I step in and out according to what I think is best for me and for the project.

HUO: The High Line was a very special collaboration with Diller Scofidio + Renfro and James Corner Field Operations, the lead architect in this project; it was also, in a way, the most public project you've been involved with. It's not just that it is world famous, but it's a little like Frank Gehry's 'Bilbao effect': every city now wants a High Line. There are projects that become …

PO: … iconic.

HUO: Yes, iconic, in that all of a sudden every city realizes what it was missing. But the High Line is also incredibly public – your most urban garden, I suppose. Can you tell us a little about how you developed it?

PO: I was working on the Battery project in New York when Field Operations, the landscape architecture firm, asked me if I would join as planting consultant for the High Line. I wasn't sure if I could do another project there at the same time, but when I looked at the website and saw so many people so passionate about it, I said yes.

From the beginning they thought I was just someone who knew a lot about plants, but I saw it as an opportunity to work in public again, to show my work in a different context and to more people. Field Operations came up with a narrative for that one and a half miles of garden, starting with woodland, opening up into a meadow, and transforming into a semi-woodland, as well as more distinct planting ideas. That narrative gave me big ideas about not just doing a garden but creating urban landscapes that could remind you of natural landscapes. From that moment I started to think completely differently about private gardens, landscape, and public gardens.

HUO: I'm going to ask you about the Hauser & Wirth garden in Somerset in a minute, but between it and the High Line is the Serpentine. Peter Zumthor, designer of the Serpentine Pavilion for 2011, asked us to recommend a landscape or garden designer, and I will always remember that first encounter between you and Peter, which Julia Peyton-Jones and I organized. He had this idea of a cloister, a monastery. That's almost the exact opposite of the High Line.

PO: Peter and I met, and I think he had already sketched his idea of what he wanted me to do. But I told him I could not do it. It was a pavilion that was quite probably in shade in the summer. So I said I had to do something that I thought would work, and not something he thought he would like. I think he appreciated this. After a discussion, he said, 'Just do what you think is right.'

I thought, this is a pavilion that has to look good from June till November. So it is, in fact, an installation, not a planting that you would do anywhere else. So I made a plan with a lot of plants, in the context of this silent building. It felt like a surprise when people came in and saw this world of plants. The first time Peter saw it, when it was just planted, he said, 'What is this plant doing here in the corner?' 'That will grow,' I said. 'It will look good in June or July.' Because he's an architect he wanted to see everything mature and ready.

HUO: You planted it in layers. You explained to me that this was necessary so that the visitor could read the confusion of leaves and stems.

PO: What I mean by 'layering' is a succession of flowering periods. It's different from what we call 'layering' in woodland, which is having a layer of trees and then a layer of perennials and one of bulbs. This was layering in seasons, so that there was something special whenever you came in. It looked completely different in June compared to September. There were plants that were coming and going.

TS: Given that we're talking about the Serpentine Gallery, I have a question in relation to art. You said you don't want to have an office. That's interesting, because lots of my artist colleagues have started to adopt the model of the architecture office. Then there are others – like Maurizio Cattelan, you and me – who want to remain more 'home alone' and be creative without a big office. Was that a conscious decision for you? I see the way you live your life as being more like an artist than an architect.

PO: I cannot have a big office because I feel distracted if I have people around me for too long. I can only do my work by myself. I can talk about my work with people, but I can only work when I'm on my own.

TS: What is your relation to art?

PO: I never had the idea of being an artist. I just wanted to be a plant person, and express myself through plants. If I can do something that touches me, and touches other people, that is my goal. That's also the drive, because if there was no recognition, I would be very unhappy.

TS: In my opinion, that's what one calls an artist.

HUO: Yes, it's a really interesting question, Tino, because gardens are increasingly recognized – your gardens particularly, Piet – as an artistic form, as creative expression. But I think in Somerset something else happened, because your drawings were celebrated there for the first time. There was a lot of discussion about colour in your work, partly because of the colour in the drawings but also because your gardens are increasingly colour-rich. Tell us about the commission in Somerset, and how you designed it.

PO: I think it's about structure more than colour, to be honest, but colour comes on top of it. Of course, I keep the colours in balance, so that they don't shine off at one end of the garden and not at the other. Repetition and balance are always part of it. But what is left after colour is structure. When you design a garden, you have to think in terms of character. How tall will a plant be? What leaf will it have? What will it do after flowering? A plant is more interesting when it makes a contribution for longer than just the flowering period. I think gardens are most interesting after flowering, because then everything becomes still and quiet, and – in my opinion – even stronger than in the flowering period.

When I choose colours for my drawings, it does not always have to do with the colour of the plants. It is more so that I can read the drawing better. Sometimes the colour yellow will not mean that a plant or flower is yellow. It sometimes comes close to the colour, but of course you run out of yellow sometimes. Then I use brown or green. It's a schematic design, so that I can quickly see what is happening.

HUO: Your drawing style changed over the 2000s, and there are now more symbols and colours.

PO: Yes, over time the drawings began to change when I started to use what they call matrix plantings as well. Block planting you see as circles and blocks in my drawings, and the design you see next to that, with little blocks and stars, is matrix planting. The latter is easier to design but more complicated to maintain, because you need a really good gardener to keep an eye on how it develops. As I've said, gardening is becoming more naturalistic, producing something that can take care of itself with editing and give the viewer the feeling that it is close to letting go, to being allowed to evolve by itself. We talked about how far we can go with gardening and still control it. Block planting is more controllable.

HUO: One thing I've felt strongly in all your gardens that I've seen is emotion. It's as much about feeling as about seeing. You once told me that it's not nature, but about bringing in an emotion of nature.

PO: Yes, it's something you feel without necessarily knowing what you see. Sometimes it's surprise. You can walk in nature and have a feeling you cannot describe, because the beauty is too large for you to understand. Maybe you can say that in gardens, too. Sometimes you cannot understand why it touches you, because it is just plants that someone has put together.

TS: Funnily enough, although I have spent my professional life surrounded with paintings, I never feel strong emotions around them; but the more of your gardens I see, the more painterly I find them. It's almost like live Impressionism.

PO: I'm not a painter, but I can see things in my mind that I can translate into the ground. I can see things that will happen after a few years. I know that what I do on paper will turn into what I have in mind. I like much art, but I don't see it as inspiration.

TS: But I do feel that the later gardens have more of this intermingling of plants, where it becomes blurry, like an Impressionist work. It depends how close you go, of course, but it seems like the more you work, the more you intermingle, or blur.

PO: Yes, but I see a lot of things photographically in the gardens. I see the beauty or the composition, but something has to happen before it is there. You cannot just put it on a canvas. I have to put it on paper first, and it has to be carried out by people, but then it has to grow. It needs a good gardener, and if you're lucky, all these things come together. If we have the right people, if they care for it, my design will look good, as I have it in my mind.

TS: I compose with people, and I think that's why I'm touched by your work. I see it as a contemporary artistic expression, because it's composing with living things. I think the twenty-first century is dealing with processes that are alive, whereas the twentieth was more about packaging processes into products. I don't say that critically, but when you talk about these details and who you work with and the dramaturgy of a garden, that's a very contemporary way of artistic working, and one that I find difficult to imagine with a straightforward sculpture or painting. To me, what you do is great visual art.

PO: I think in time and processes. In time, because in my work you have to. If I plant an oak tree, I need to think about where I plant it, because it will become very big. When I work with a shrub, I know that in five years' time it will be that size. Thinking in processes also involves thinking about what can happen if the garden is neglected. The trees will stay, and the shrubs. So there must be a grounding of every garden I do that will stay after the play of the perennials, which I see as ephemeral. It might stay for twenty years, but then it feels tired, and you want to do something different. You can see the garden as a stage play in which plants perform, but also as something that creates ideas for others.

TS: I would say it's an immersive play, because you can walk in it. You can see it from further away, or you can see it from close up.

PO: In your own life, when you look back, things haven't all happened at one time. The development of how I think at work is the result of forty years.

HUO: That brings us to something that many artists at the moment are fascinated by in your work and in gardens today more generally. For example, the artist Precious Okoyomon has created a garden in Aspen, Colorado. Instead of an exhibition, the artist decided to make a garden. The art world is still very occupied with exhibitions, all connected to events, but everybody is now aware that we need to think in longer, more sustainable terms.

PO: Gardens will never die in people's ideals, as they see the world. You can see nature as the biggest garden we have, and we must treasure it, but we also want to create our own worlds, and that is what the garden will be. But I see gardens as ephemeral. Everything I do is a playful intermezzo with time, and what I like or what other people like will be different in twenty years. Gardens will change, and our idea of them will change. The idea of how a garden will look - the beauty and the aesthetics - is always based in the time in which we live.

TS: I have a last question for you, Piet. Earlier, you and Hans Ulrich were talking about the High Line 'landscape'. What is your favourite landscape at the moment?

PO: There's no favourite, because every landscape is its own world. When I go to where I lived when I was younger, in the western Netherlands, walking or cycling through the pine trees and the dunes and coming to the sea, that reminds me of my childhood. I think the best landscapes are in your memory.

HUO: That leads me to my last question, about unrealized projects. Architects' unrealized competition projects are published and enter the public consciousness, but artists or writers or philosophers or landscape architects don't publish in that sense. You don't, because you don't do competitions, normally. Obviously, the scope of the unrealized is broad. There are projects that are too big to be realized, too expensive or too time-intensive, projects that are censored, and projects we don't dare to do - self-censorship. Can you tell us about one or two of your favourite unrealized projects, your dreams?

PO: I have no idea. I can tell you that when they called me for the High Line, nobody had any idea what it would become. I have been offered bigger projects that I know I could not realize, because time is an issue at my age. I want a project to be ready within five years, not ten.

Who would have said that Hauser & Wirth Somerset would become such a great garden among my portfolio, if I may say that? You cannot anticipate. You have to believe in your client first, as I did with Iwan and Manuela Wirth, and with you, Hans Ulrich, at the Serpentine. It is the client who makes you believe in a project. I need a client first, to make my work, whereas an artist usually needs to make work first and then find a client.

Tino Sehgal is an artist of German and Indian descent working in Berlin. His work takes the form of live encounters between visitors and those enacting the work, and has been presented in museums and galleries around the world.

Museo Chillida Leku
Hernani, Spain
2019

Chillida Leku is a museum created by the Basque sculptor Eduardo Chillida to display a range of his work. An 11 hectare (27 acre) park of trees, mown grass and lush greenery provides a location for some 40 steel and concrete abstract forms, and is a place where, in his words, people 'could rest and where people could walk through them as though in a forest'. A traditional Basque farmhouse is a museum for drawings, maquettes and smaller works.

Opened originally in 2000, the museum started a collaboration with Swiss art dealers Hauser & Wirth and re-opened in 2019 after a prolonged closure. Piet has been involved in creating a small area of perennial planting around the entrance, as well as a long tree and shrub border at the rear of the parking area that makes the most of the very good growing conditions in this part of the world – abundant year-round rainfall and mild winters. Azaleas, magnolias, dogwoods and lilacs are included, continuing a theme of planting established by Chillida himself.

Late summer in the Chillida Leku garden. The tall pink perennial at the rear is *Eupatorium maculatum* 'Riesenschirm', very popular with butterflies, and a classic 'back of the border' plant. *Overleaf*: A planting plan for a perennial border (top left) and (below left) a shrub and small tree planting by the parking area. The perennial border uses blocks with *Amsonia hubrichtii* and *Phlomis tuberosa* 'Prima Donna' as scatter plants, putting in a distinct rhythm earlier in the season than the majority of the components.

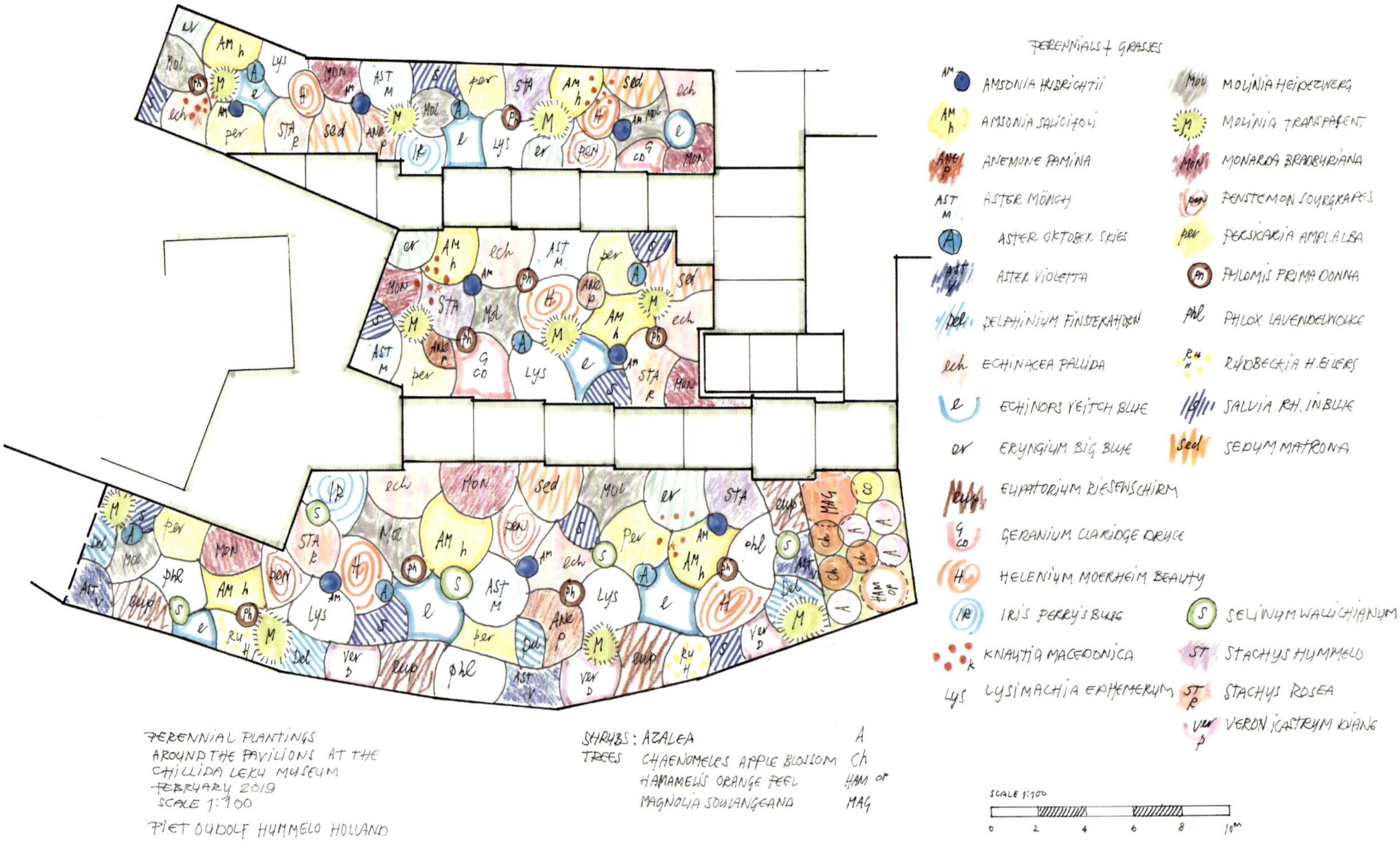
PERENNIALS + GRASSES
AMSONIA HUBRICHTII
AMSONIA SALICIFOLIA
ANEMONE PAMINA
ASTER MÖNCH
ASTER OKTOBER SKIES
ASTER VIOLETTA
DELPHINIUM FINSTERAHORN
ECHINACEA PALLIDA
ECHINOPS VEITCH BLUE
ERYNGIUM BIG BLUE
EUPATORIUM RIESENSCHIRM
GERANIUM CLARIDGE DRUCE
HELENIUM MOERHEIM BEAUTY
IRIS PERRY'S BLUE
KNAUTIA MACEDONICA
LYSIMACHIA EPHEMERUM
MOLINIA HEIDEBRAUT
MOLINIA TRANSPARENT
MONARDA BRADBURIANA
PENSTEMON SOURGRAPES
PERSICARIA AMPL. ALBA
PHLOMIS PRIMA DONNA
PHLOX LAVENDELWOLKE
RUDBECKIA H. EILERS
SALVIA RH. IN BLUE
SEDUM MATRONA
SELINUM WALLICHIANUM
STACHYS HUMMELO
STACHYS ROSEA
VERONICASTRUM DIANE
PERENNIAL PLANTINGS
AROUND THE PAVILIONS AT THE
CHILLIDA LEKU MUSEUM
FEBRUARY 2019
SCALE 1:100
PIET OUDOLF HUMMELO HOLLAND
SHRUBS: AZALEA A
TREES CHAENOMELES APPLE BLOSSOM Ch
HAMAMELIS ORANGE PEEL HAM OP
MAGNOLIA SOULANGEANA MAG
SCALE 1:100
0 2 4 6 8 10m

TREES AND SHRUBS
AESCULUS PARVIFLORA
AMALANCHIER LAMARCKII MS
AZALEA JAPONICA
AZALEA MOLLIS
CALLICARPA BODINIERI
CERCIS FOREST PANSY
CHAENOMELES APPLE BLOSSOM
CHAENOMELES NIVALIS
CHOISYA AZTEC
CLERODENDRON BUNGEI
CLERODENDRON TRICHOTOMUM
CORNUS STELLAR PINK
CORNUS KOUSA CHINA GIRL
CORNUS MAS
CORYLOPSIS PAUCIFLORA
DAPHNE BURKWOODII
GLEDITSIA INERMIS / MS
HALESIA MONTICOLA
HAMAMELIS ARNOLD PROMISE
HAMAMELIS JELENA
HAMAMELIS ORANGE PEEL
HAMAMELIS DIANE
HYDRANGEA VILLOSA
LAURUS NOBILIS
MAGNOLIA SOULANGEANA
PAEONIA ROCKII + OTHER
PHILADELPHUS CORONARIUS
PRUNUS YEDOENSIS
PTELEA TRIFOLIATA
SYRINGA MISS KIM
VIBURNUM BODNANTENSE DAWN
PLANTING PLAN FOR THE PARKING AREA
AT CHILLIDA LEKU SAN SEBASTIAN
PIET OUDOLF HUMMELO HOLLAND
SCALE 1:100
FEBRUARY 2019
SCALE 1:100
0 2 4 6 8 10m

Block planting at Chillida Leku: orange *Helenium* 'Moerheim Beauty' and yellow *Rudbeckia fulgida* prominent in the entrance planting (above top), and *Sanguisorba* 'Blackthorn' (above).

Private Garden
Basel, Switzerland
2020

Part of this project is a roof garden – one large area designed for access and two smaller areas intended to be seen from above only – but Piet also designed perimeter plantings on the lower, ground level. Roof gardens are a key feature of his design portfolio, as can be seen with the Lurie Garden in Chicago (see pp.74–81) and the High Line in New York (see pp.82–95). Private clients in dense urban areas are inevitably turning to every available space to garden, even if not all areas are normally accessible. There are also huge environmental advantages to roof greening projects, as they absorb stormwater and release it slowly back to the atmosphere or ground drainage.

Here the substrate is approximately 30 centimetres (12 in) deep, and additional irrigation is available if needed to ensure plants survive during extremely dry periods; there are a few notably drought-tolerant species (e.g. *Geranium sanguineum* and *Asclepias tuberosa*), but otherwise the plant mix makes few concessions to having reduced moisture reserves in the soil. A few birch trees (well adapted to roof gardens because of their shallow roots) add height and additional winter interest with their white trunks. An outer shaded area on a slope uses a woodland ground flora.

Two bold clumps of *Aster* 'New Jersey Blue Skies' are part of the late-season look for this roof garden. The narrow, mosaic stone path is entirely appropriate to the intimacy of a private garden but would not be seen in any larger projects.

Left: Sketch for the main area of the roof garden. *Opposite, top:* The roof garden at eye level. *Opposite, bottom:* A shaded area adjoining the neighbouring property, with *Liriope* 'Ingwersen' – a grass-like plant very useful for ground cover – prominent. New planting also took in the perimeter of the building, and inevitably ground and roof level share species and overall feel.

Opposite, top: A *Liquidambar* tree in a narrow border in the ground is a major focus for autumn; the image below it is another view of the main area of the roof garden: stepping stones are used to provide a minimal path, and note the absence of planting immediately around the birch trunks – compromises like this are inevitably part of the challenge of creating gardens on roofs. *Above*: Final plan for the main roof garden; a cultivar of the relatively small prairie grass *Schizachrium scoparium* is used as a scatter plant for summer to winter interest.

Oudolf Garten on the Vitra Campus
Weil am Rhein, Germany
2020

Located in southwest Germany within walking distance of the Swiss border, the Vitra Campus is the historic production site of the furniture manufacturer Vitra. Commissioned by Vitra's chairman emeritus, Rolf Fehlbaum, several notable contemporary architects have been commissioned to design buildings for the various facilities here, including Álvaro Siza, Zaha Hadid and Frank Gehry. In 2018 Piet was commissioned to create a 3,600 square metre (38,750 sq. ft) perennial planting.

Such a large area of relatively uniform cultural conditions offers many opportunities for developing planting combinations using different proportions of the palette. The main area of planting is given definition by a mature cherry tree and two grass viewing mounts. Piet's design also took into consideration two sculptural interventions called *Ring* and *Ruisseau*, which were designed by Ronan and Erwan Bouroullec in 2018. *Ring* is a circular steel bench positioned under the cherry tree, while *Ruisseau* is a line of marble featuring a narrow channel filled with running water.

Framed and screened by hedging, the centrepieces of Piet's work here are two areas of matrix planting, using *Sporobolus heterolepis* for one and *Sesleria autumnalis* for the other. Within the matrix there are perennial blocks; some perennials are planted as individuals for structure (e.g. *Baptisia* varieties) and others as short-season scatter plants (e.g. cultivars of *Papaver orientale*). The outer areas are screening plantings using taller perennials. Among these is a small percentage of blocks where two to four varieties are included - typically to provide two seasons of interest from plants that co-exist well or, alternatively, two cultivars of the same species. Other smaller areas around the VitraHaus building by Herzog & de Meuron use matrix and block plantings repeating the same plant palette. The garden was planted in 2020 by a team of professional horticulturalists, overseen by Bettina Jaugstetter in Piet's absence (given Covid-19 travel restrictions).

View over the central area of the Vitra Campus in mid summer (this picture was taken in June). The bed in the foreground uses block planting; the four areas around the tree use a matrix.

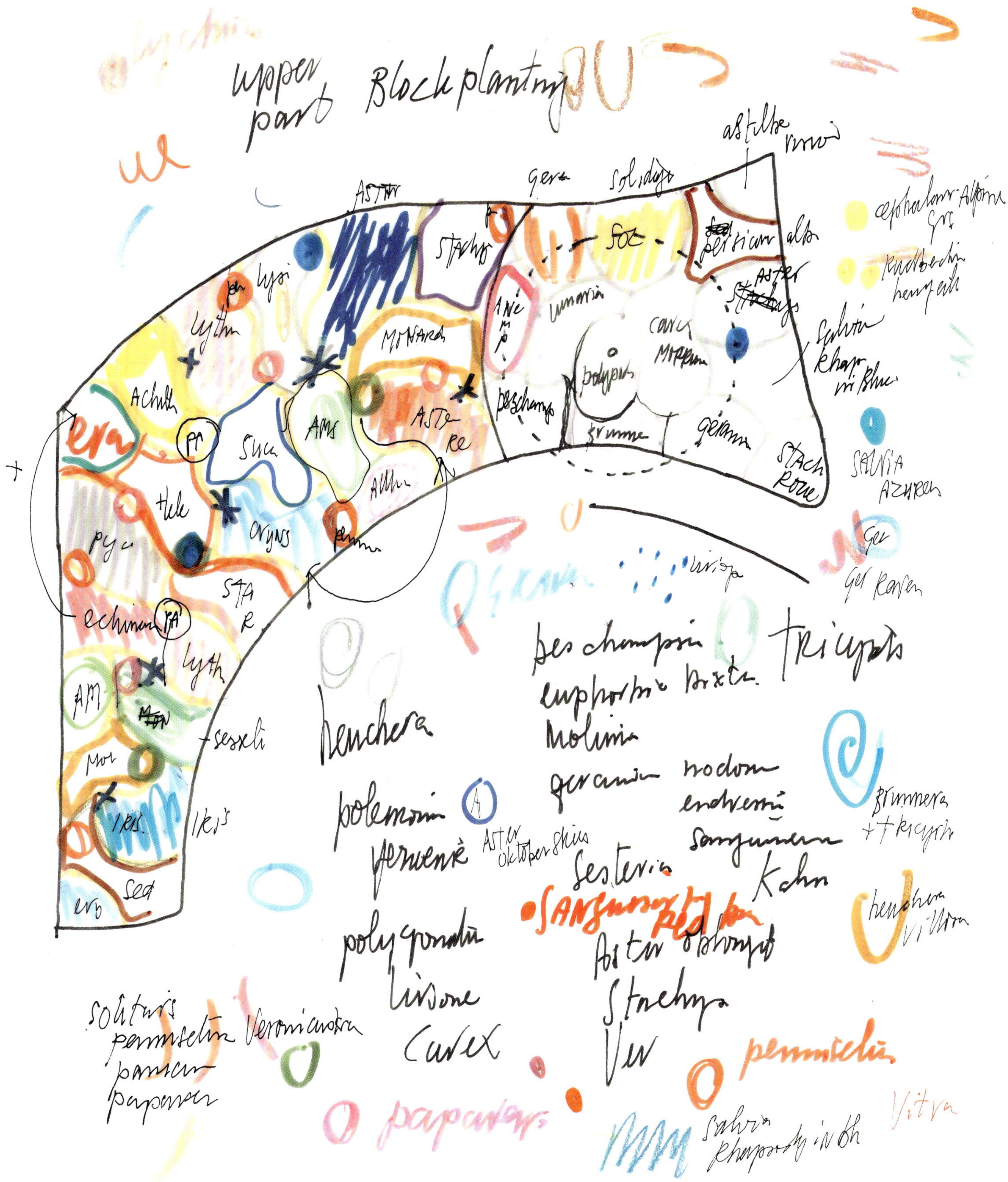

Preliminary sketches (also overleaf) and detail of one of the areas of block planting in mid summer in July: deep pink *Echinacea* 'Fatal Attraction', violet *Perovskia* 'Little Spire' and orange *Helenium* 'Moerheim Beauty'. The contrast between block and the central matrix planting encourages viewers to see similar plants in very different contexts.

Bloch planting

Text OMDRAAIEN
Bloch planting
Veronica

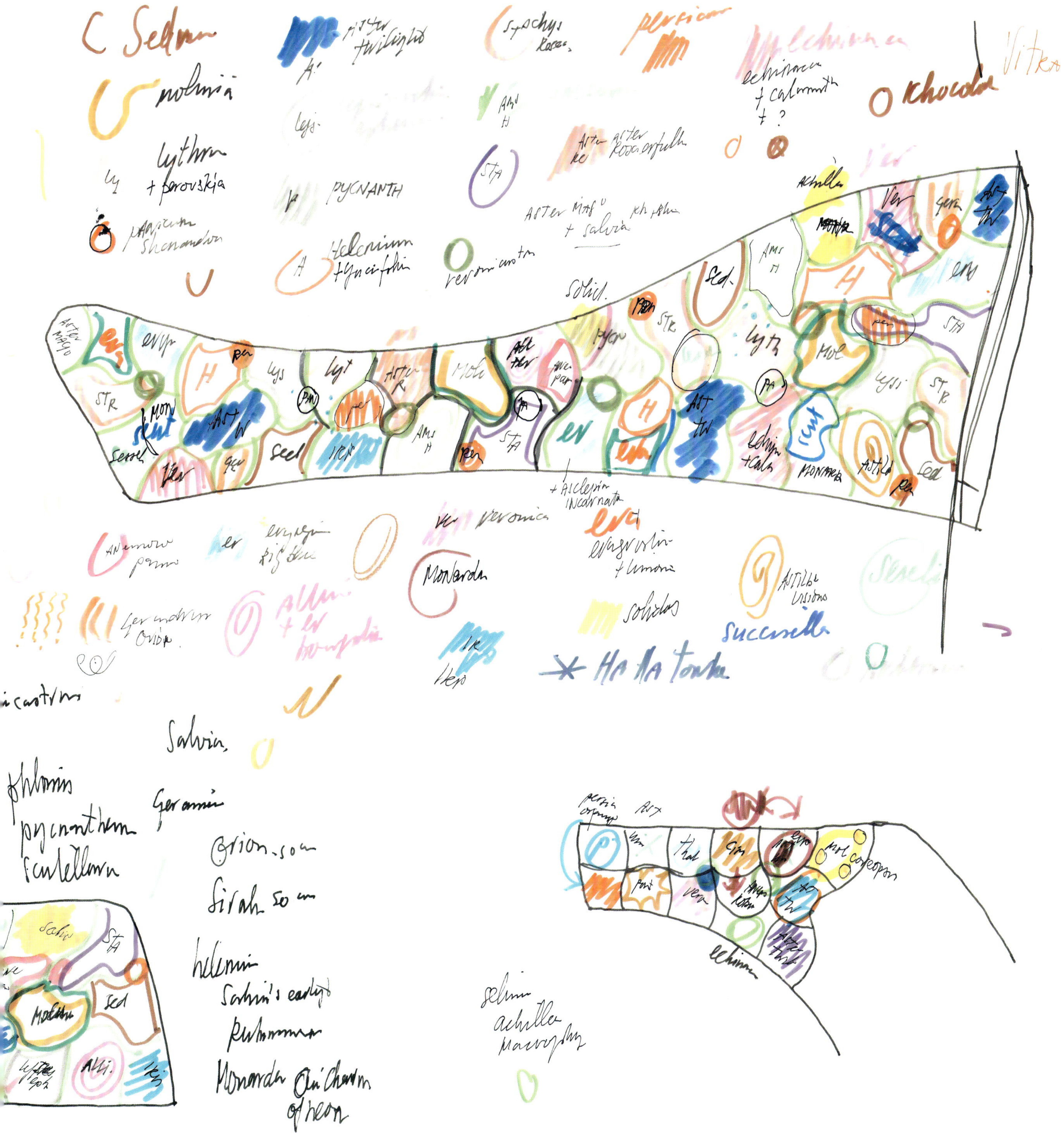

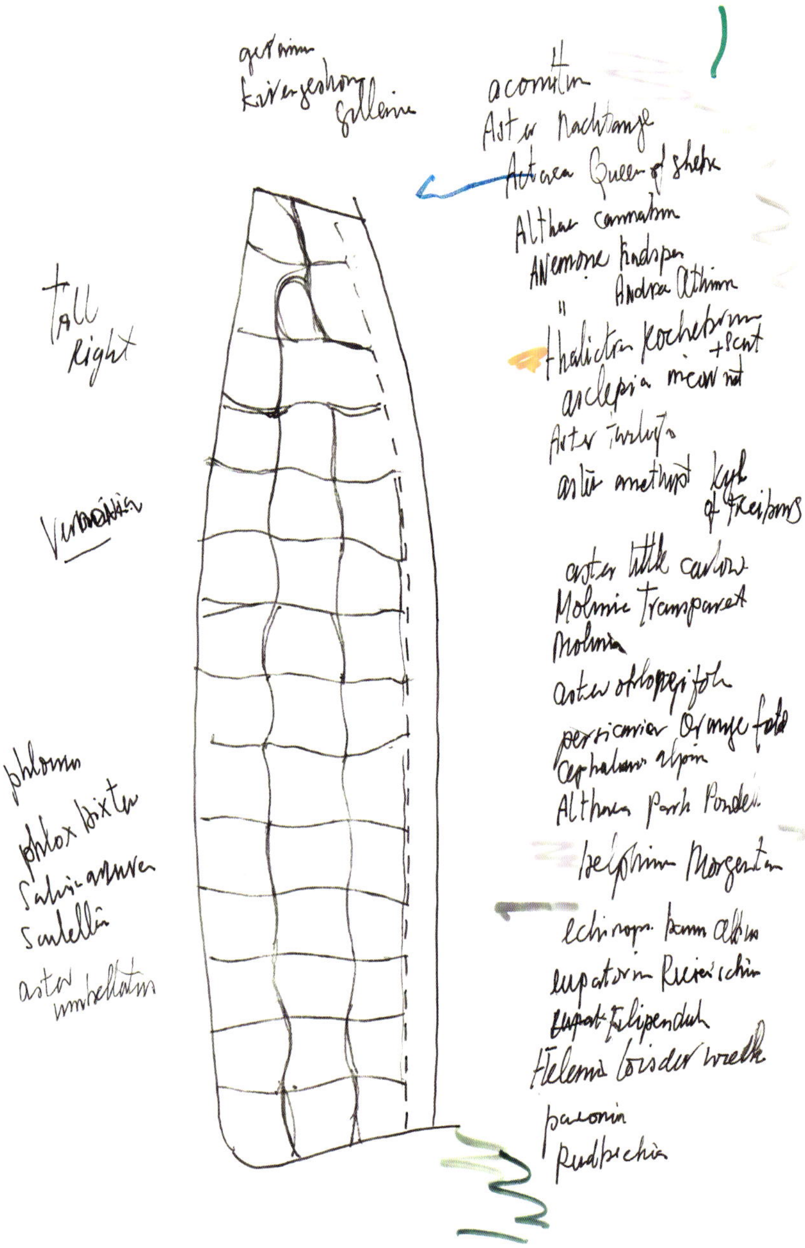

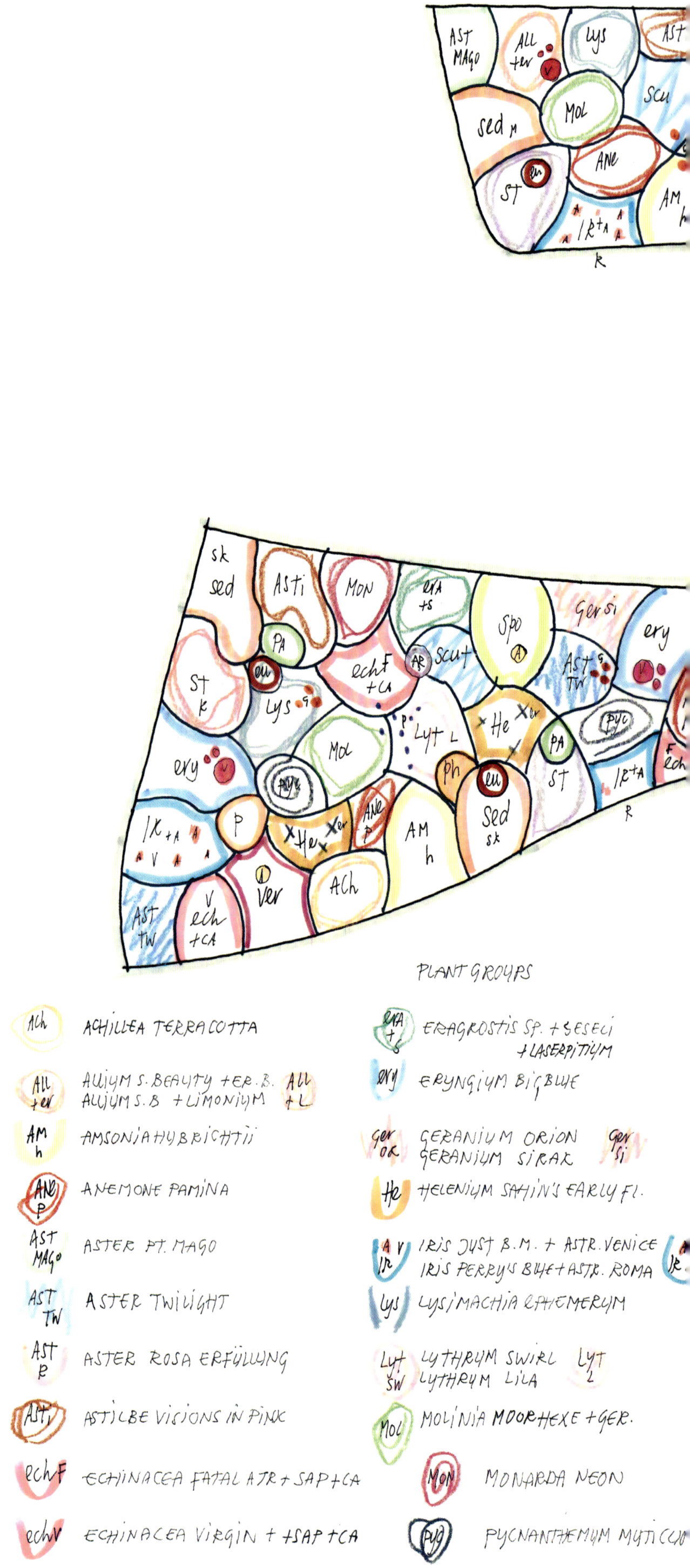

Finished planting plan with a preliminary sketch, where plant juxtapositions and distribution are planned out.

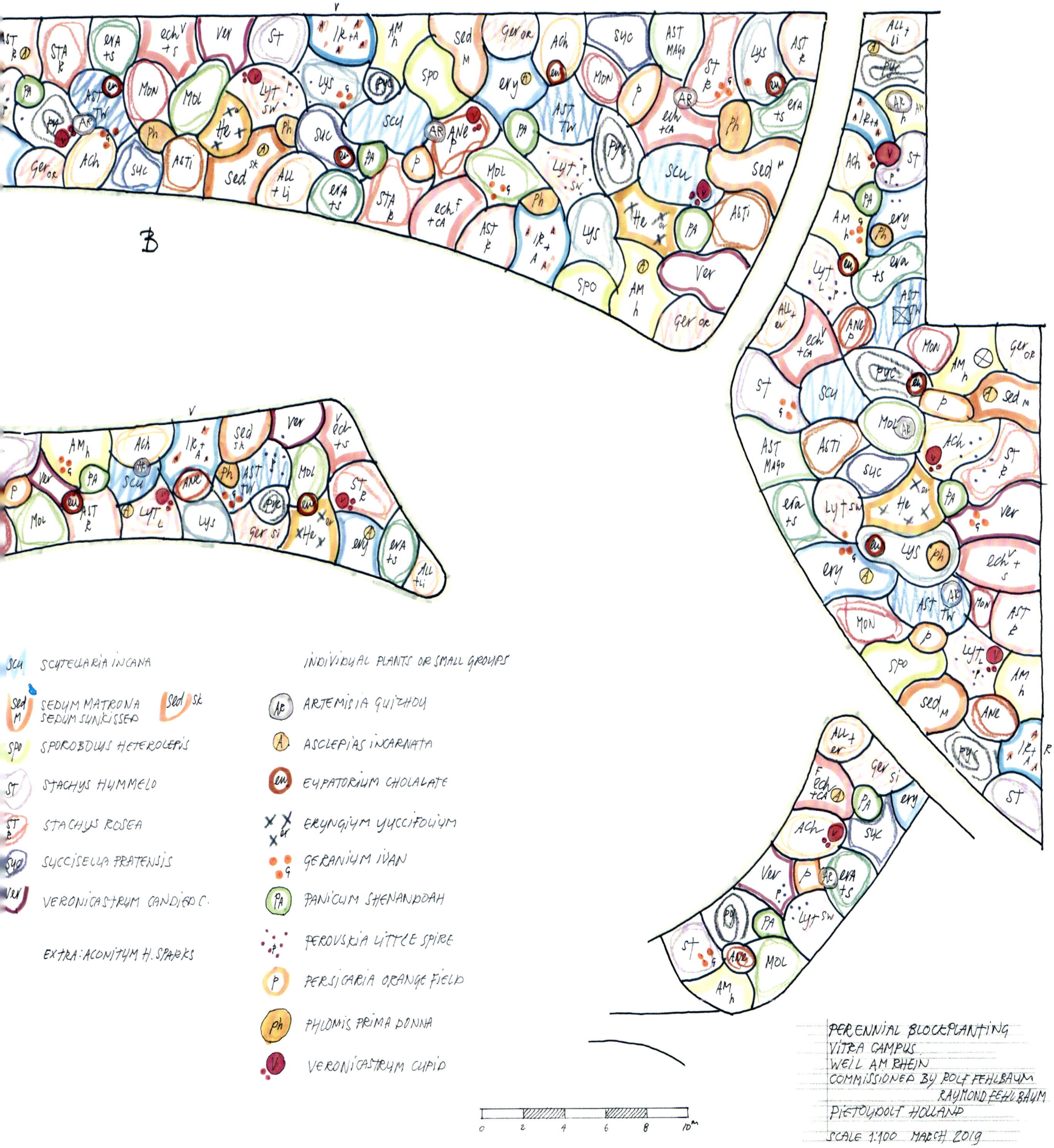
B
SCU SCUTELLARIA INCANA
SED M SEDUM MATRONA
SED SK SEDUM SUNKISSED
SPO SPOROBOLUS HETEROLEPIS
ST STACHYS HUMMELO
ST R STACHYS ROSEA
SUC SUCCISELLA PRATENSIS
VER VERONICASTRUM CANDIED C.
EXTRA: ACONITUM H. SPARKS
INDIVIDUAL PLANTS OR SMALL GROUPS
AR ARTEMISIA GUIZHOU
A ASCLEPIAS INCARNATA
EU EUPATORIUM CHOCOLATE
ER ERYNGIUM YUCCIFOLIUM
G GERANIUM IVAN
PA PANICUM SHENANDOAH
P PEROVSKIA LITTLE SPIRE
P PERSICARIA ORANGE FIELD
PH PHLOMIS PRIMA DONNA
V VERONICASTRUM CUPID
0 2 4 6 8 10 m
PERENNIAL BLOCKPLANTING
VITRA CAMPUS
WEIL AM RHEIN
COMMISSIONED BY ROLF FEHLBAUM
RAYMOND FEHLBAUM
PIET OUDOLF HOLLAND
SCALE 1:100 MARCH 2019

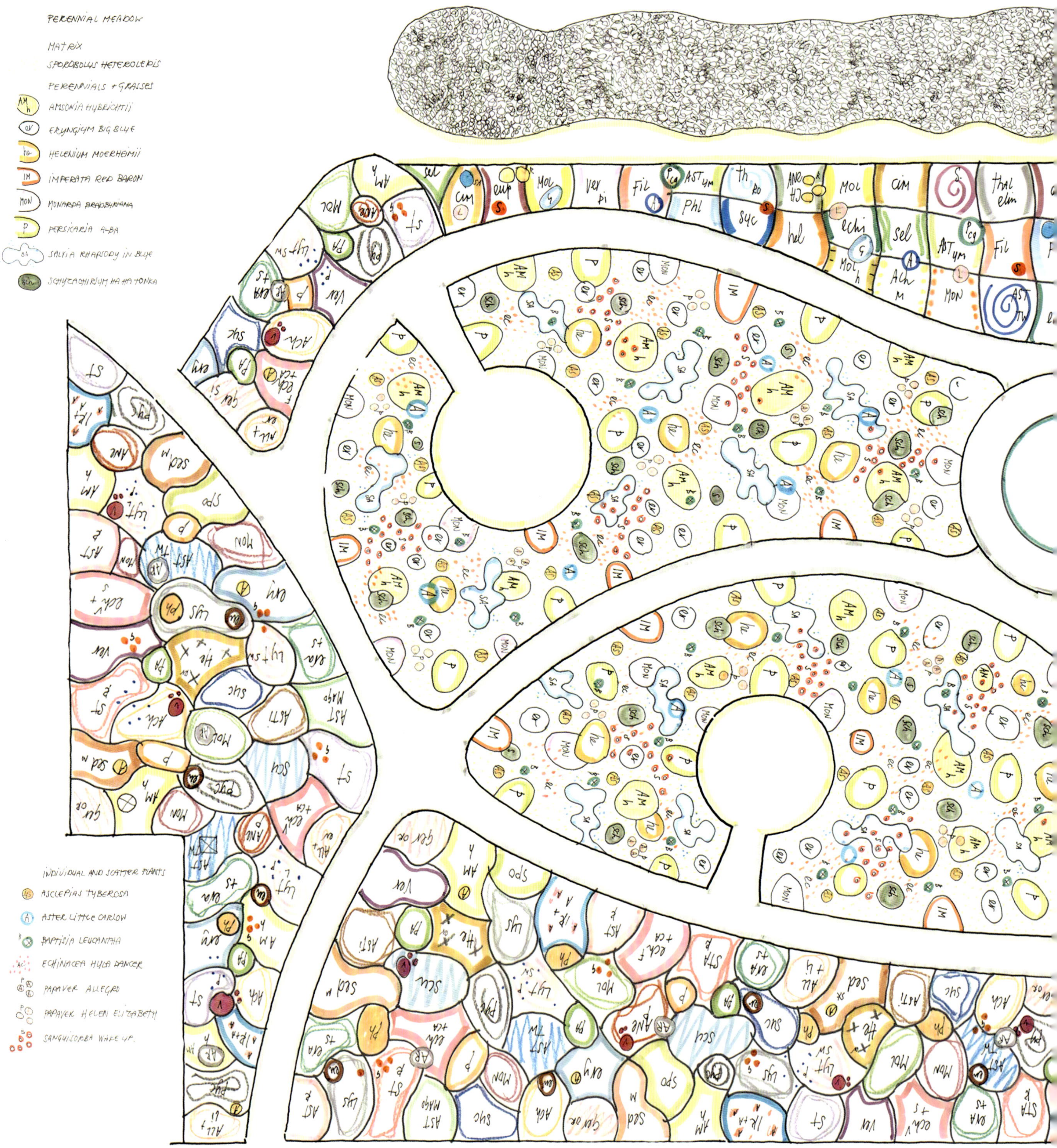

The central area of the garden in spring (previous pages), a plan of the central area (above) and the garden in full flower in late summer (overleaf). *Ring* and *Ruisseau* (2018) by Ronan and Erwan Bouroullec are in the centre and to the right of the design. The big clumps of strongly colourful perennials around the outside can be seen as framing the more subtle matrix plantings in the centre, a roughly similar arrangement to the Oudolf Field at Hauser & Wirth in Somerset (see pp.134–35). Taller perennials in the outer rim of planting are backed by hedging; creating a frame for large-scale plantings is important to help define their limits and therefore focus attention on what is designed to be looked at.

Clockwise from upper left: The centre of the campus planting in early summer, with the geodesic dome which is used for events; yellow *Eremurus* varieties make a powerful impact at the beginning of the main flowering season; as summer advances, seedheads begin to take over from flowers; *Cenolophium denudatum*; grasses make a major contribution later in the season.

Late-season perennial and grass plantings remind us of the oft-quoted Oudolf phrase about a plant only 'being worth growing if it looks good in winter'.

Private Garden

Nieuw Haamstede, Zeeland, The Netherlands
2020–21

This house and garden in the southwestern Netherlands uses mostly deciduous trees and shrubs to screen and give context to three areas of roof garden, including many native species, such as alder and beech, as well as flowering shrubs like rhododendron and magnolia varieties, very much a mix of the traditional and the less common. Some areas also have a ground flora of shade-loving plants and spring bulbs. From the roof, the boundaries are difficult to see, and the property's planting merges with the greenery of the surroundings. The green roof substrate is 20–25 centimetres (8–10 in) deep, on the low side for strong growth of many perennials, and the plant selection reflects this, with several species whose ancestry is that of shallow, dry soils, such as *Origanum laevigatum* 'Hopley's Variety' and *Salvia* × *sylvestris* 'Blauhügel'. There is irrigation available, however. Individually placed structural plants include the drought-tolerant grass *Festuca mairei* and the bright-flowered *Geum triflorum*.

The shortage of garden space increasingly drives property owners to develop green roofs and, if they have a good designer, to integrate these with the land that they have on the ground and possibly even the surroundings. With more and more people turning to high-density living, we can expect this to be an important design style of the future.

Rooftop planting is a golden opportunity to use European native steppe and dry meadow species, all of them tolerant of drought conditions and thin soil. Here the brief season of the spectacular wispy grass *Stipa barbata* hovers over varieties of lavender and *Salvia nemorosa* 'Blauhügel' and 'Rhapsody in Blue', with pale yellow *Achillea* 'Moonshine'.

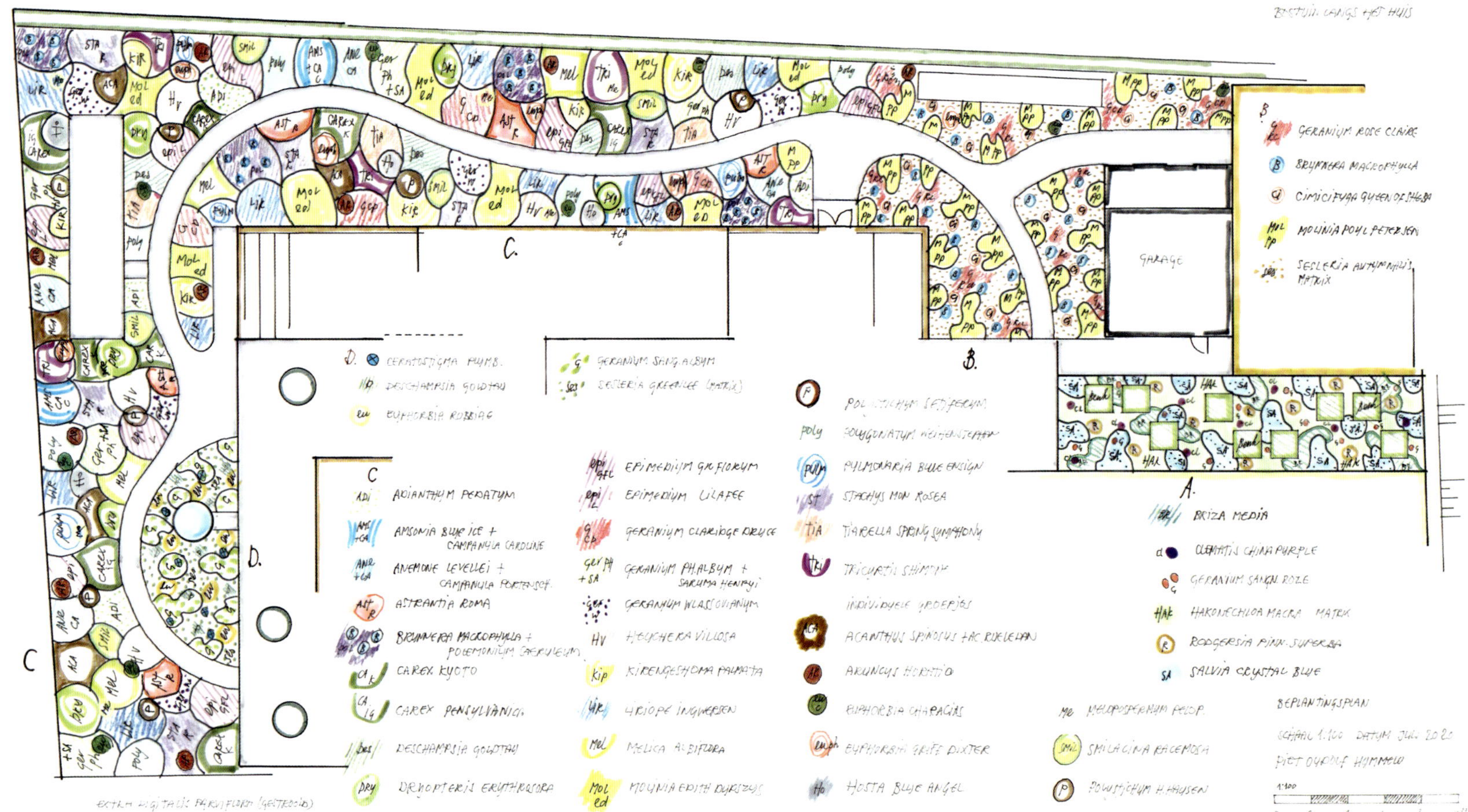

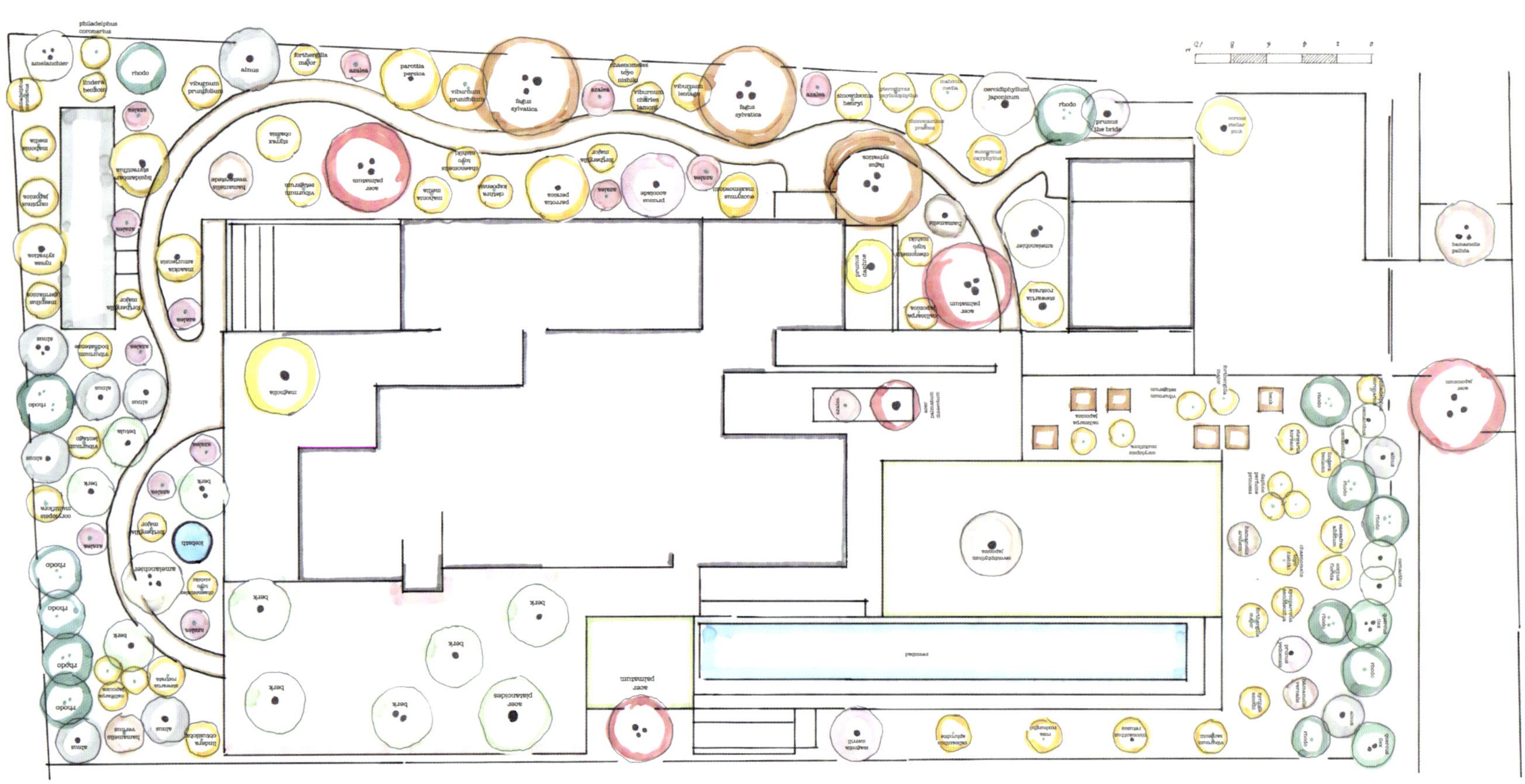

Plan for underplanting the outer rim of trees and shrubs with woodland perennials (top left), and one of several plans for the tree and shrub planting (bottom left). Shown opposite are plans for the roof gardens.

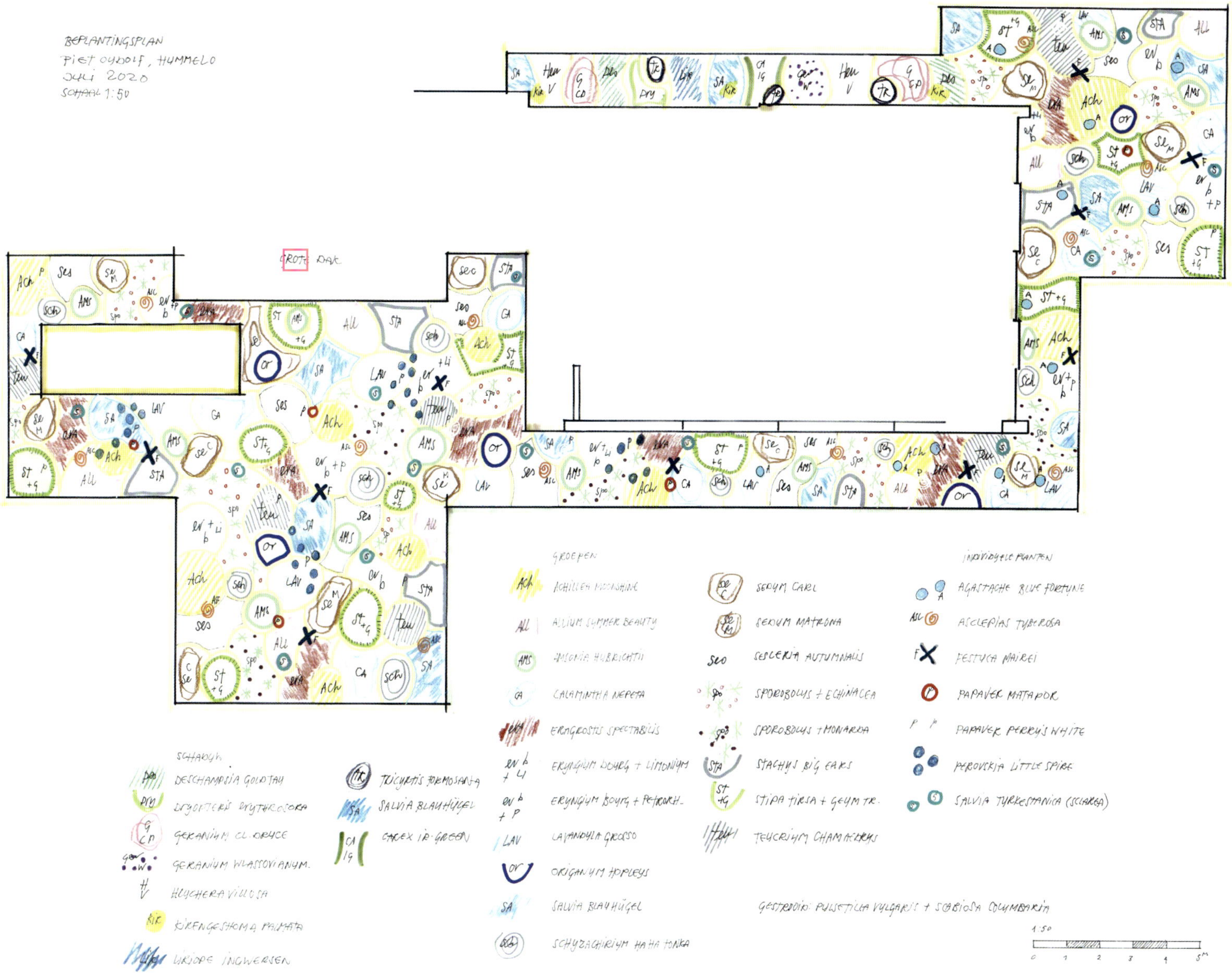
BEPLANTINGSPLAN
PIET OUDOLF, HUMMELO
JULI 2020
SCHAAL 1:50
GROEPEN
ACHILLEA MOONSHINE
ALLIUM SUMMER BEAUTY
AMSONIA HUBRICHTII
CALAMINTHA NEPETA
ERAGROSTIS SPECTABILIS
ERYNGIUM BOURG + LIMONIUM
ERYNGIUM BOURG + PEROVSK.
LAVANDULA GROSSO
ORIGANUM HOPLEYS
SALVIA BLAUHÜGEL
SCHYZACHIRIUM HA HA TONKA
SEDUM CARL
SEDUM MATRONA
SESLERIA AUTUMNALIS
SPOROBOLUS + ECHINACEA
SPOROBOLUS + MONARDA
STACHYS BIG EARS
STIPA TIRSA + GEUM TR.
TEUCRIUM CHAMAEDRYS
INDIVIDUELE PLANTEN
AGASTACHE BLUE FORTUNE
ASCLEPIAS TUBEROSA
FESTUCA MAIREI
PAPAVER MATADOR
PAPAVER PERRY'S WHITE
PEROVSKIA LITTLE SPIRE
SALVIA TURKESTANICA (SCLAREA)
SCHADUW
DESCHAMPSIA GOLDTAU
DRYOPTERIS ERYTHROSORA
GERANIUM CL. DRUCE
GERANIUM WLASSOVIANUM
HEUCHERA VILLOSA
KIRENGESHOMA PALMATA
LIRIOPE INGWERSEN
TRICYRTIS FORMOSANA
SALVIA BLAUHÜGEL
CAREX IR. GREEN
GESTROOID: PULSATILLA VULGARIS + SCABIOSA COLUMBARIA
1:50

BEPLANTINGSPLAN
PIET OUDOLF, HUMMELO
JULI 2020
SCALE 1:50
VASTE PLANTEN GROEPEN
ALLIUM SUMMER BEAUTY
ANEMONE CYLINDRICA
CALAMINTHA NEPETA
COREOPSIS MOONBEAM
DALEA PURPUREA + ARMERIA
ERAGROSTIS SPECTABILIS
ERYNGIUM BOURGATII + LIMONIUM
FESTUCA AMETHYSTINA
ORIGANUM HOPLEYS
SALVIA BLAUHÜGEL / RH. IN BLUE
STACHYS BIG EARS
SEDUM CARL + RED CAULI
INDIVIDUELE PLANTEN
FESTUCA MAIREI
GEUM TRIFLORUM
PAPAVER MATADOR
PAPAVER PERRY'S WHITE
SALVIA TURKESTANICA
SPOROBOLUS HETEROLEPIS
EXTRA: PULSATILLA VULGARIS

Planting here makes the most of the disparity between harsh conditions on the roof and the lushness of light shade. *Acanthus mollis* 'Rue Ledan' makes a strong architectural statement (left), while North American *Echinacea pallida* joins a mostly European native plant selection on the roof (top, centre). Perennials frame a formal pool (top right) and woodland species much of the periphery of the property (above).

Oudolf Garden on Belle Isle
Detroit, Michigan, USA
2021

This 1.2 hectare (3 acre) garden sits in front of the Nancy Brown Peace Carillon tower on Belle Isle on the Detroit River, a historic public space now run as a Michigan state park, and Piet's design comprises 15 main beds, a rain garden and a wetlands garden/habitat. The garden came about when members of the Garden Club of Michigan invited Piet to design a project for them on a site adjacent to the historic Anna Scripps Whitcomb Conservatory. The garden continues to be run and managed by a volunteer group: Oudolf Garden Detroit.

Belle Isle was originally laid out by Frederik Law Olmsted, designer of New York's Central Park, and includes an early twentieth-century greenhouse and botanical garden at the conservatory. The site is low-lying and areas are prone to flooding, so Piet's design consists of a raised section planted with perennials that stays dry when water levels rise. Here the design of the planting beds was inspired by the flow of the water in the original terrain and the beds feature a variety of block and matrix plantings. Another small area was designed as a rain garden, where water can sit temporarily after heavy rainfall before soaking gently away. The rain garden concept is one that is finding rapid take-up as part of SUDS (Sustainable Urban Drainage Schemes), and the range of perennial species that can be used is quite wide.

A view over Oudolf Garden on Belle Isle from the belltower shows how low-lying this site is. The orientation of the beds reference the patterns in the soil on the original site, created by the flow of water. The planted area is raised with a dyke to the left of the path.

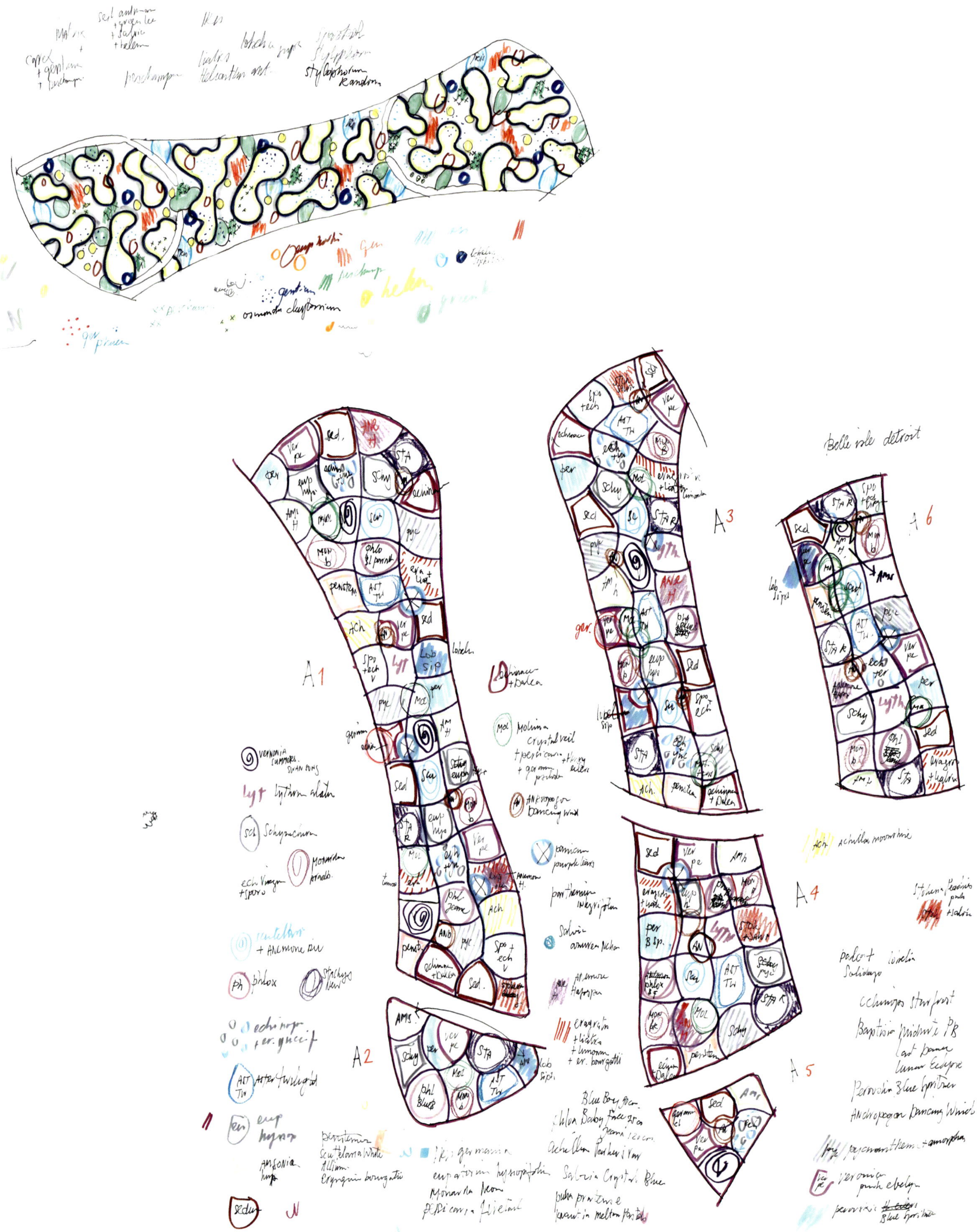

Preliminary sketches of planting beds. The one at top left is the rain garden bed, which involved a matrix of *Sesleria autumnalis*, *Carex albicans* and *Carex montana*.

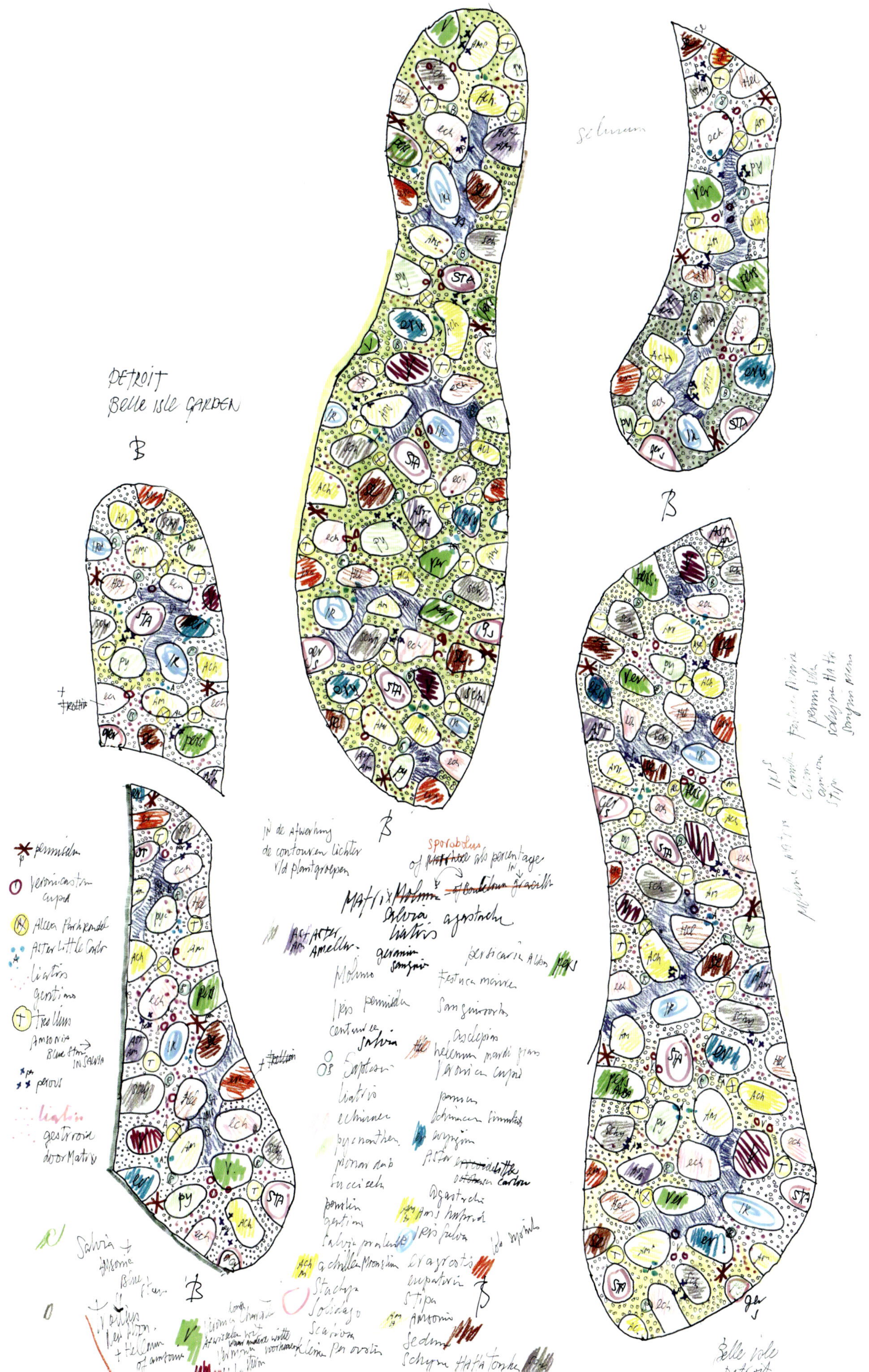
DETROIT
BELLE ISLE GARDEN

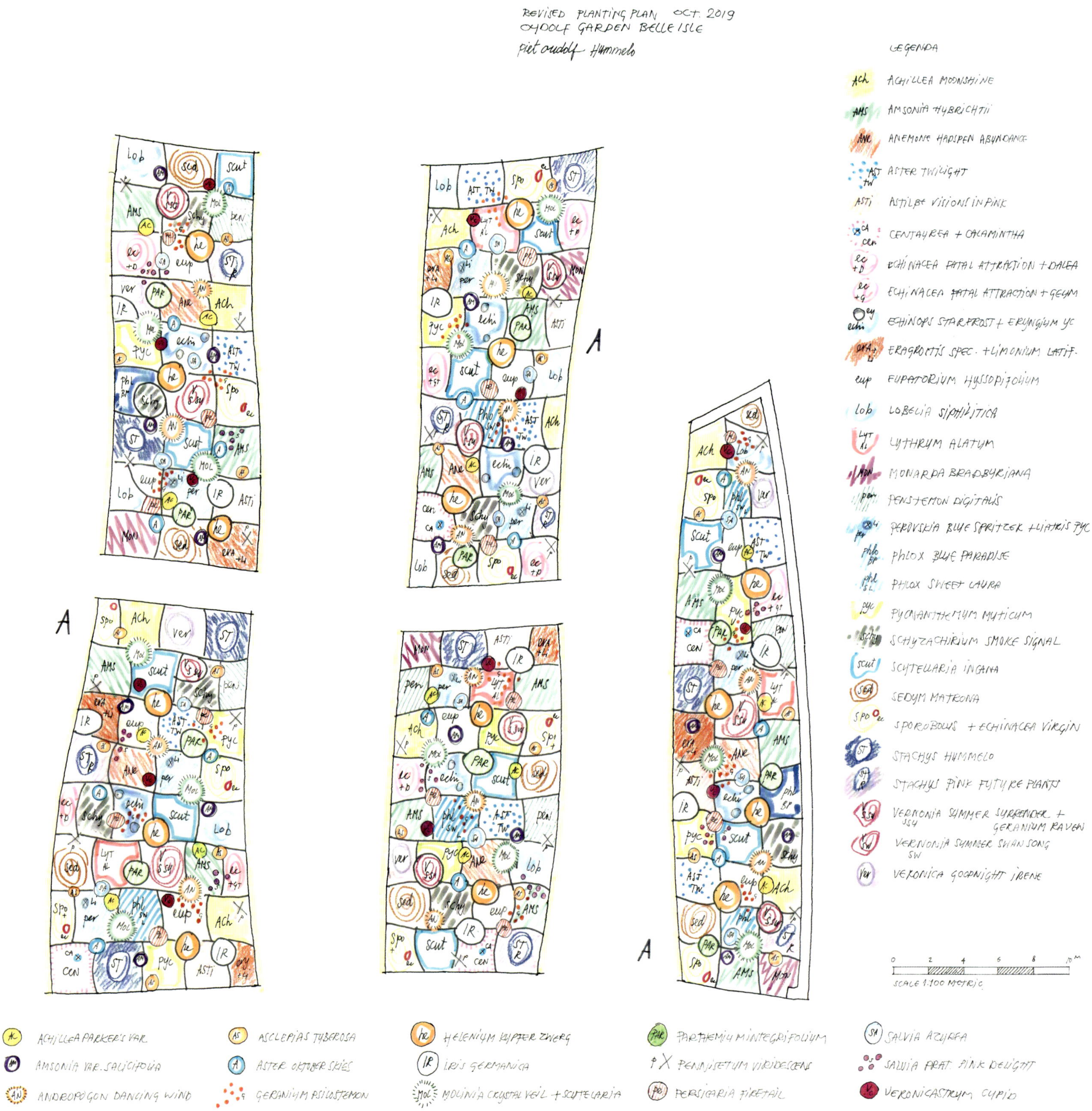

Planting plans for the Belle Isle project. A perennial planting is above and to the right is a bulb plan, including *Allium*, *Puschkinia* and *Tulipa* varieties, alongside the native woodland perennial *Mertensia virginica*.

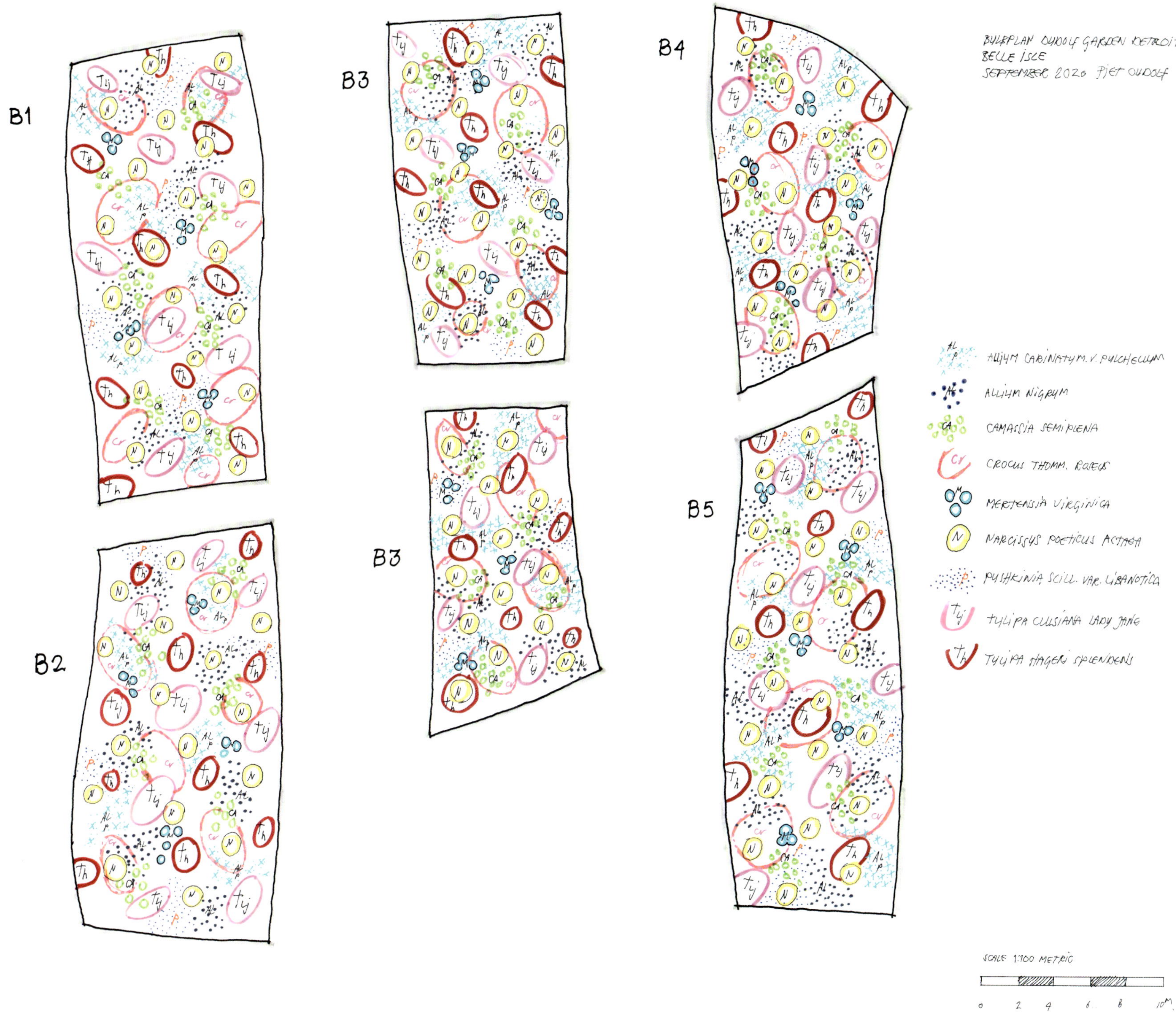

BULBPLAN OUDOLF GARDEN DETROIT
BELLE ISLE
SEPTEMBER 2020 PIET OUDOLF
B1
B2
B3
B3
B4
B5
ALLIUM CARINATUM V. PULCHELLUM
ALLIUM NIGRUM
CAMASSIA SEMIPLENA
CROCUS THOMM. ROSEUS
MERTENSIA VIRGINICA
NARCISSUS POETICUS ACTAEA
PUSHKINIA SCILL. VAR. LIBANOTICA
TULIPA CLUSIANA LADY JANE
TULIPA HAGERI SPLENDENS
SCALE 1:100 METRIC
0 2 4 6 8 10M

A view down the central axis of the garden in summer looking east towards the early twentieth-century Anna Scripps Whitcomb Conservatory, with details of the planting – with camassia in spring (above top); *Deschampsia cespitosa* 'Goldtau' and *Lobelia siphilitica* in the rain garden (above centre); and *Liatris pycnostachya* at sunset (above right).

Oudolf Garden at Hauser & Wirth Menorca
Illa del Rei, Menorca, Spain
2021

A Hauser & Wirth Gallery on the Illa del Rei in Mahon harbour, Menorca, incorporates a number of outdoor exhibition areas and includes a small garden alongside the gallery buildings, which have been converted from an eighteenth-century naval hospital. The gallery's permanent collection includes works by Eduardo Chillida and Joan Miró, and the gallery presents exhibitions and has an education centre. This is one of the few examples of an Oudolf garden outside a temperate zone climate; summers here are hot and dry, while the growing season runs from the first rains in autumn until early June.

The plant selection is radically different from what Piet normally uses. All the plants chosen will tolerate summer drought, at least once they are established. Most are Mediterranean region natives: a mixture of perennials, sub-shrubs (small, compact, evergreen woody plants) and grasses. Notable non-natives include *Melianthus major*, a dramatic large-leaved perennial; *Gaura lindheimeri*, a North American perennial that has become widely popular because of its remarkable tolerance of heat and drought; and the American grass *Muhlenbergia capillata*.

Mediterranean planting with *Elogio del Vacío VI* (2000) by Eduardo Chillida. *Overleaf*: Grass *Eragrostis spectabilis* and grey-leaved *Artemisia* 'Powis Castle' make a strong impact – in zones with dry summers, evergreen or long-lasting foliage inevitably plays a larger role than in more humid climates; flowering times are short and even seedhead interest may be curtailed. The planting perfectly complements the sculpture *Le Père Ubu* (1974) by Joan Miró.

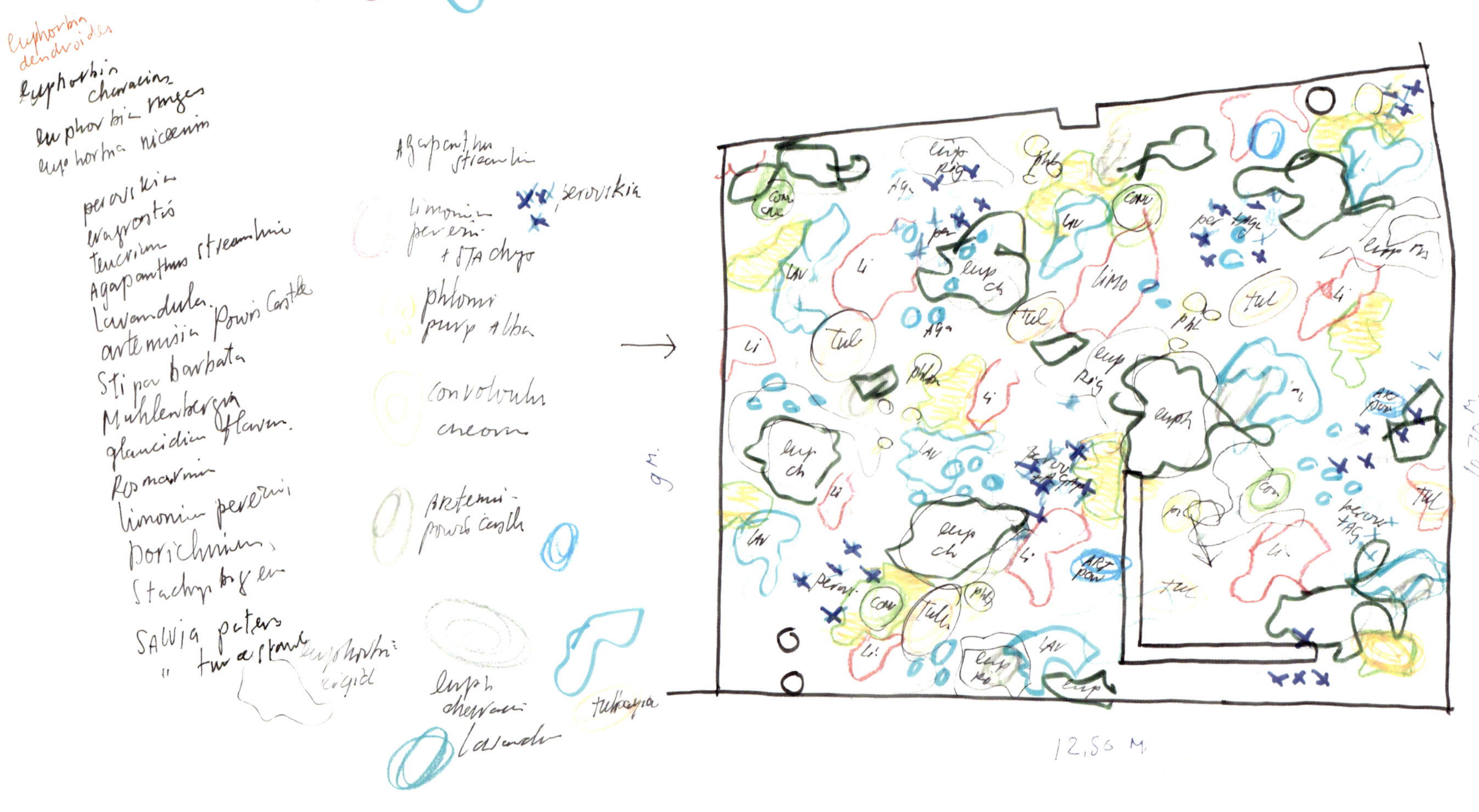

Sketches (above) and finished planting plans (opposite). The grass *Eragrostis spectabilis* is used very much as an infill, being a relatively low-growing species with very fine flower- and seedheads that often complement more defined plant forms.

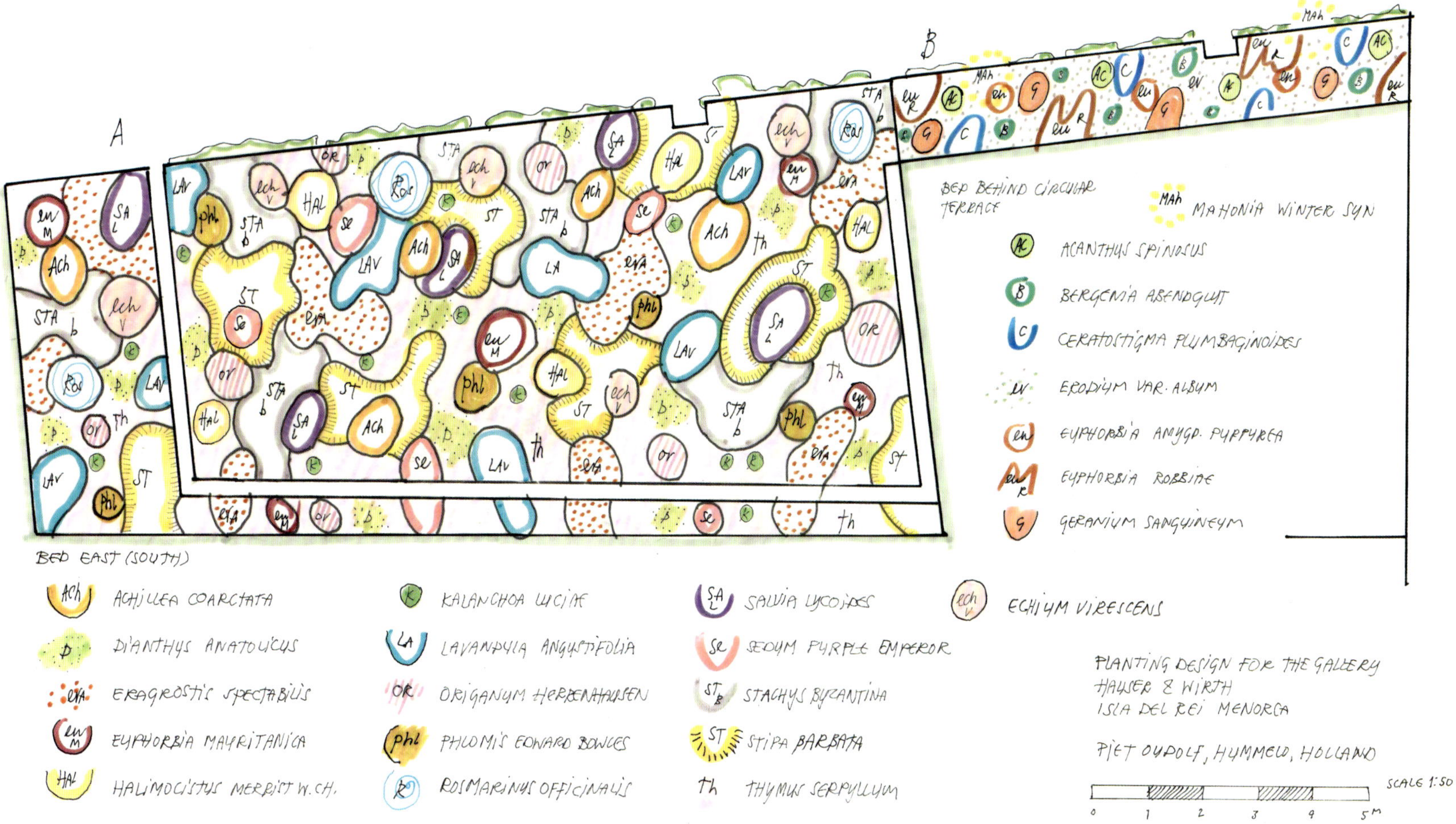

A
B
BED BEHIND CIRCULAR TERRACE
MAHONIA WINTER SUN
ACANTHUS SPINOSUS
BERGENIA ABENDGLUT
CERATOSTIGMA PLUMBAGINOIDES
ERODIUM VAR. ALBUM
EUPHORBIA AMYGD. PURPUREA
EUPHORBIA ROBBIAE
GERANIUM SANGUINEUM
BED EAST (SOUTH)
ACHILLEA COARCTATA
DIANTHUS ANATOLICUS
ERAGROSTIS SPECTABILIS
EUPHORBIA MAURITANICA
HALIMOCISTUS MERRIST W.CH.
KALANCHOA LUCIAE
LAVANDULA ANGUSTIFOLIA
ORIGANUM HERRENHAUSEN
PHLOMIS EDWARD BOWLES
ROSMARINUS OFFICINALIS
SALVIA LYCOIDES
SEDUM PURPLE EMPEROR
STACHYS BYZANTINA
STIPA BARBATA
THYMUS SERPYLLUM
ECHIUM VIRESCENS
PLANTING DESIGN FOR THE GALLERY
HAUSER & WIRTH
ISLA DEL REI MENORCA
PIET OUDOLF, HUMMELO, HOLLAND
SCALE 1:50
0 1 2 3 4 5M

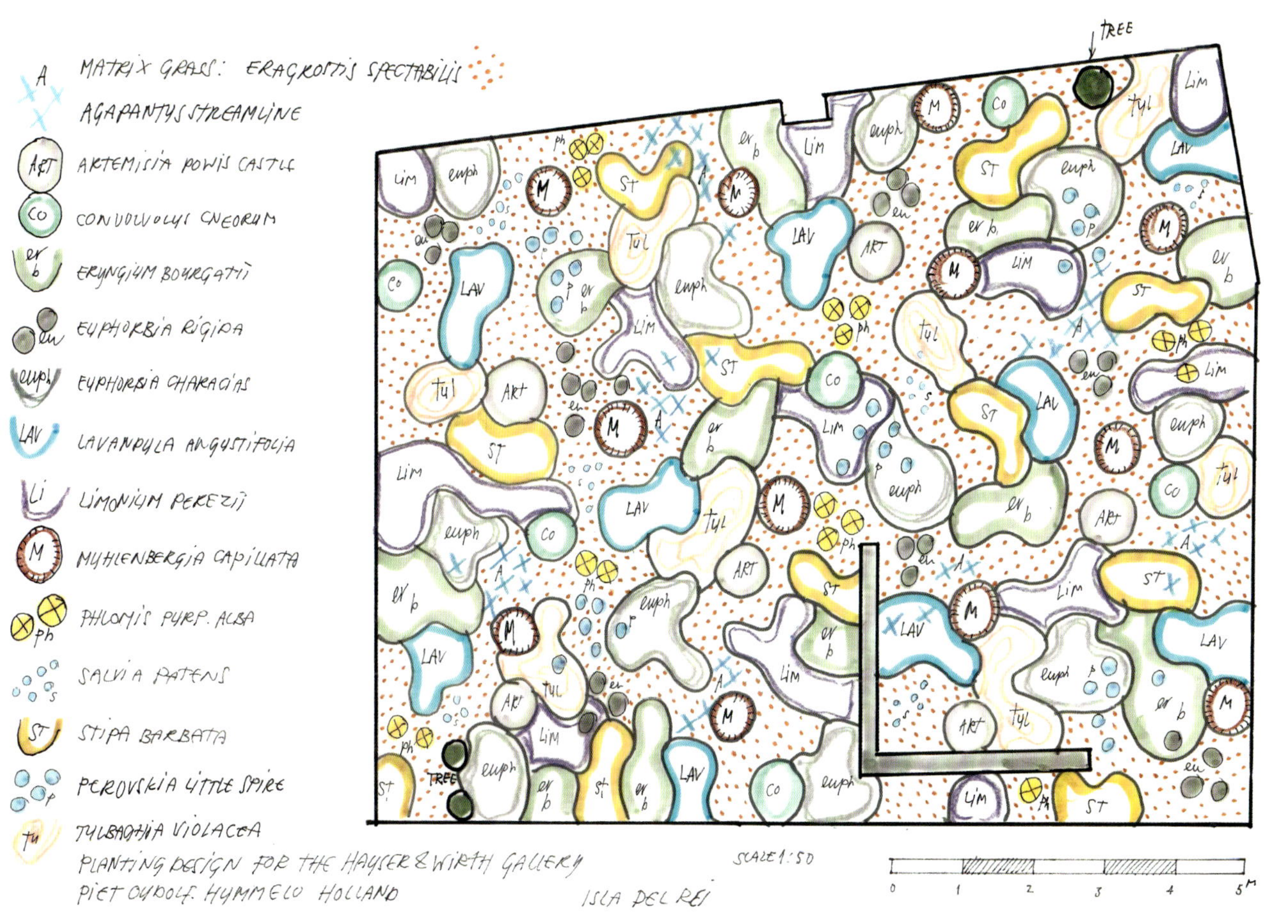

MATRIX GRASS: ERAGROSTIS SPECTABILIS
AGAPANTUS STREAMLINE
ARTEMISIA POWIS CASTLE
CONVOLVULUS CNEORUM
ERYNGIUM BOURGATTI
EUPHORBIA RIGIDA
EUPHORBIA CHARACIAS
LAVANDULA ANGUSTIFOLIA
LIMONIUM PEREZII
MUHLENBERGIA CAPILLARIS
PHLOMIS PURP. ALBA
SALVIA PATENS
STIPA BARBATA
PEROVSKIA LITTLE SPIRE
TULBAGHIA VIOLACEA
PLANTING DESIGN FOR THE HAUSER & WIRTH GALLERY
PIET OUDOLF HUMMELO HOLLAND
ISLA DEL REI
SCALE 1:50
TREE
0 1 2 3 4 5M

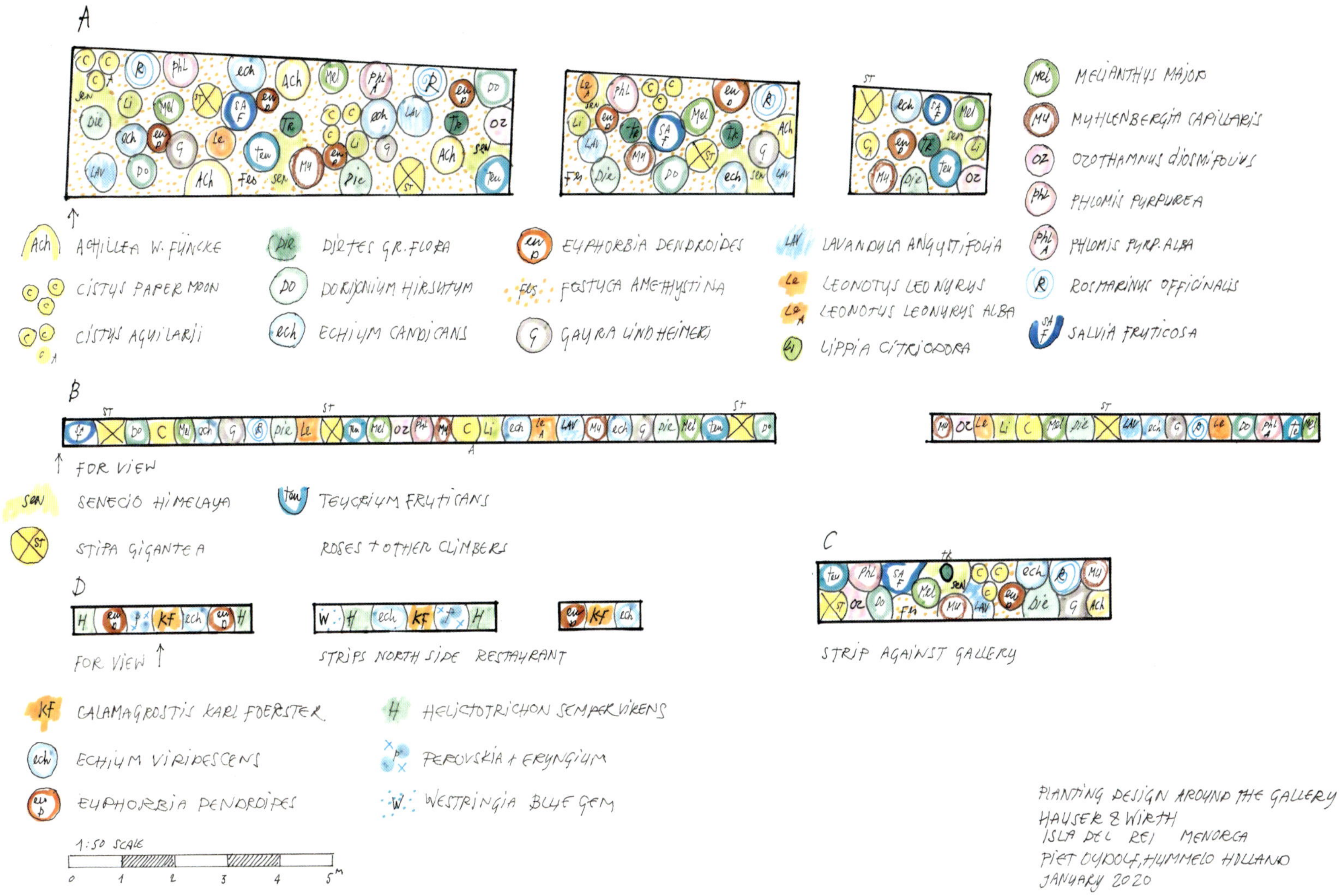

Planting complements the old stone walls of the former naval hospital – the eighteenth-century buildings were restored and house the gallery spaces. *Above:* Planting plans for a number of small spaces around the gallery buildings. In one of these the low-growing grass *Festuca amethystina* (not pictured) is used as a matrix.

Plants around the gallery buildings illustrate the wide range of growth forms to be found among drought-tolerant plants. *Echium candicans* (lower left) is a good example of a species whose form and long season of distinctive seedheads make it a good Oudolf plant. *Leonotius leonurus* (above centre) has a distinctive form, with whorled flower clusters superficially similar to those of the *Monardas* that appear in much of Piet's work in moister climates.

CANTINA

I think the best
landscapes
are in your memory.

Piet Oudolf

The Evolution of a Plantsman

Rosie Atkins

Piet Oudolf's career is defined by what he calls his 'healthy obsession with plants', and the plant knowledge he has accumulated over the years is the key to his ground-breaking designs.

Born on 27 October 1944, Piet grew up in the natural landscape around Bloemendaal in the Netherlands [1], where his parents ran a busy bar and restaurant. An adventurous youth, he remembers clambering over their garden fence to explore the neighbouring fields, woodlands and expanse of dunes that stretch to the North Sea, an area that is now Zuid Kennemerland National Park. Half a mile from Bloemendaal is Thijsse's Hof, one of Europe's oldest wildlife reserves, created by pioneer ecologist and botanist Dr Jacobus P. Thijsse (1865–1945). In 1925 Thijsse was gifted a garden of 0.8 hectares (2 acres) for his birthday, which he rewilded using native plants from a specialist nursery in nearby Overveen. Thijsse's Hof is the inspiration for what the Dutch call *heemparks* – areas of wilderness that provide retreats from city life. All these early influences would inform the approach of the plantsman Piet became.

Like most young boys, Piet took the wild beauty of his surroundings for granted, but he acknowledges that it has had a lasting effect on his appreciation of nature. He was in his early twenties when he realized the family business wasn't for him, and instead found work in a local plant nursery, where he began propagating plants. This experience and books about gardening led him to study landscape construction and contracting, which he pursued with a view to setting up his own garden design and maintenance business.

In 1968 Piet met Anja, and they married in 1970, buying a little house in Haarlem, just west of Amsterdam. As soon as their sons, Pieter and Hugo, were old enough, Piet and Anja would take the boys on plant-hunting expeditions to England. Travelling Dutch-style, by bicycle, they visited the gardens and nurseries of such plant pioneers as Alan Bloom (1906–2005) at Bressingham, Norfolk; Christopher Lloyd (1921–2006) at Great Dixter in Sussex; and Beth Chatto (1923–2018) at her eponymous gardens in Essex. They returned to Haarlem determined to find somewhere with the space to experiment with plants and expand Piet's business.

Finally, in 1982, Piet and Anja bought a derelict farm outside the village of Hummelo, in the eastern Netherlands near the German border, with 1.3 hectares (3.2 acres) of land [2, 3]. They soon realized they would need help setting up the nursery and went into partnership with Romke van der Kaa, who had been head gardener at Great Dixter in the 1970s. A few years later, Van der Kaa left to start his own nursery and pursue what became a successful career as a gardening writer and broadcaster. Piet's garden design business subsequently went on hold while he established the nursery, which was becoming known for its perennials and unusual grasses. For inspiration, Piet and Anja would visit the German gardens and nurseries of such famous plantsmen as Karl Foerster (1874–1970) and Georg Arends (1863–1952), and they became good friends with the influential plant breeder Ernst Pagels (1913–2007), who was a key figure in Piet's development.

In 1983, the year after the family moved to Hummelo, Henk Gerritsen (1948–2008) and his partner, Anton Schlepers (1945–1993), visited the nursery [9], struck by a plant list that contained things unavailable in the Netherlands at the time. Gerritsen, who was to become a lifelong friend, was a trained artist and keen botanist, and he and Schlepers were creating the Priona Gardens at their home in Schuinesloot, not far from Hummelo. Piet describes Gerritsen as 'a master of observation. He knew where plants came from in nature, which allowed him to create gardens that felt as if you were returning to the real roots of gardening and which made you realize the extent to which our relationship with nature had degenerated.' To Piet and Gerritsen, dead plants and seedheads were not ugly but beautiful, home to insects and beneficial to birds. Their ideal was the exact opposite of the traditional, high-maintenance garden.

In 1984 Piet and Anja invited Rob Leopold (1942–2005), co-founder of the wildflower seed company Cruydt-Hoeck, to take part in their first open day,

1

2

3

4

5

6

[1] Jacob van Ruisdael, *The Bleaching Grounds of Fam. De Mol near Haarlem*, c.1660, oil on canvas; [2] Aerial view of Hummelo, *c.*1982; [3] Piet and Anja at Hummelo; [4] Trial beds in the nursery at Hummelo, 1980s; [5] Presentation of the publication *Droomplanten*, 1990; [6] Piet's gold medal-winning design at Floriade garden festival, 1992.

along with other growers. The open days at Kwekerij Oudolf ('Nursery Oudolf') were not only an opportunity for nurseries to sell plants but also a chance to meet like-minded people and explore new ways of looking at plants and gardens. Late-night discussions with Leopold and Gerritsen fuelled Piet's interest in selecting plants that weren't just decorative but had architectural impact, providing year-round interest while being resistant to pests. Leopold called it 'creative ecology and integrated landscape design'.

By now the nursery at Hummelo had become the testbed that Piet envisaged would underpin his garden design practice, and the trial beds had started to yield some horticultural treasures [4]. Piet spotted and cultivated a particularly striking seedling of purple sage (*Salvia verticillata*), which he called 'Purple Rain', and a *Gaura lindheimeri* that lived up to its name, 'Whirling Butterflies'. These early selections were to become widely available commercially when he joined a collective of growers focused on breeding reliable, beautiful perennials that were also beneficial to wildlife. Later, Piet was to become a founder member of a Dutch-based global network of growers and breeders called Future Plants Ltd, set up in the 1990s to promote and protect new plant cultivars.

Kwekerij Oudolf's open days were a showcase for plants, attracting gardeners both amateur and professional, and soon became a magnet for journalists and photographers. While these events helped put Hummelo on the horticultural map, they took up a lot of Anja's time, and Piet was now in demand as a garden designer. In 1989 they reduced the open days to one weekend in autumn, calling it *Grassendagen* (Grass Days) to promote their range of ornamental grasses.

I first met Piet in the spring of 1989, when I accepted an invitation to visit the stylish biannual Journées des Plantes, held in the grounds of the seventeenth-century Chateau de Courson, just outside Paris. The show, which grew out of a local plant fair, had become a horticultural sensation, and Piet was an early member of the jury of plant experts. The British plantsman and author Roy Lancaster, a fellow member of the jury, introduced me to Piet and Anja, who invited me to the Grass Days that September. There at Hummelo, Piet and Gerritsen were excited to be working together on their first book, *Droomplanten*, which would be published in 1990 (and in 2003 in English as *Planting the Natural Garden*) [5]. *Droomplanten* first introduced Piet's plant selections to the wider world.

In 1992 Piet designed a gold medal-winning exhibit for a consortium of growers at the Dutch horticultural expo Floriade, an event that takes place every ten years [6]. He also accepted an invitation from the agronomist and dendrologist Rune Bengtsson to attend the MOVIUM conference on planting design in Alnarp, Sweden. Brita von Schoenaich, a Kew Diploma student with a PhD in Landscape Design, whom Piet had met at MOVIUM, had begun planning a symposium at the Royal Botanic Gardens, Kew, called 'Designing with Perennials: New Trends in Planting'. In 1994 she invited Piet to attend the first in what became a series of conferences that launched the New Perennial movement, celebrating a new style of planting design focused primarily on perennials.

Piet continued to broaden his knowledge and his circle of contacts. In 1995 he attended the first Perennial Perspective symposium, held at the now established annual garden show Freisinger Gartentage in Freising, Germany, where he had the good fortune to meet Stefan Mattson, Director of Parks in Enköping, Sweden. *Droomplanten* had been well received in Sweden, and their conversation led to Piet's first foreign commission, a public space that became known as *Drömparken* or 'Dream Park' [8], after the book's Swedish title, *Drömplantor*. The project allowed Piet to use many plants grown at Hummelo that were unavailable in Sweden. His 'River of Sage', using *Salvia nemorosa* 'Ostfriesland', 'Blue Hill' and 'Rugen', caused a sensation when the park opened in 1996, turning Piet and Anja into local celebrities.

Piet and Anja attended the 'Perennials in Public Green Spaces' symposium in 1996, in Arnhem, The Netherlands; and, also that year, John Coke asked Piet to design a garden in the farmyard of his home at Bury Court, near Farnham, Surrey (p.49). This was Piet's first commission in the UK, celebrating his distinctive use of perennials mixed with topiary. The following year, in 1997, they participated in 'New Trends in Planting' at the Royal Botanic Gardens, Kew. The latter gave voice to a group of American plantsmen including Neil Diboll, an expert in native plant ecology.

By now, Piet's reputation as a garden designer was gaining international recognition, and he was commissioned to design a Millennium Garden at Pensthorpe Waterfowl Park in Norfolk, in the east of England. He received a further boost when *Méér Droomplanten*, the follow-up to *Droomplanten*, was published in 1999, followed by an English translation titled *Dreamplants for the Natural Garden* in 2000. Also in 1999, Piet and Anja travelled to Slovenia with Henk Gerritsen, my husband, Eric, and me, following a recommendation by Jelena de Belder (1925–2003), the botanist and a fellow Courson jury member who had grown up there. Stane Sušnik, a well-known Slovenian gardening journalist, was our guide. We botanized in the mountains and admired the chemical-free meadows where farmers still used scythes to cut the hay. That same year, the Royal Horticultural Society asked Piet to design the double borders leading to the new glasshouse at their headquarters at Wisley, in Surrey [7].

Since meeting in 1998, Piet and the talented young British designer Arne Maynard had been working on plans for a Chelsea show garden sponsored by *Gardens Illustrated* magazine, which was launched in 1992 (I was the founding editor). Their design, called 'Evolution', was selected for the main avenue at the RHS Chelsea Flower Show in May 2000, which was an achievement in itself for two Chelsea unknowns. Piet and Maynard collaborated with Chris and Toby Marchant of Orchard Dene Nurseries, and Mark Fane and his construction team at Waterers. To our surprise, the garden won a Gold Medal and 'Best in Show', which was beyond our wildest dreams. This inevitably resulted in Piet and Maynard being inundated with requests for garden designs, and the Hummelo nursery became busier than ever, with visitors arriving by the coachload.

Around this time, Piet was asked to enter a competition to produce the planting plan for a public garden in Chicago, called the Lurie Garden (see pp.74–81). He collaborated with Kathryn Gustafson of Gustafson Guthrie Nichol, a leading firm of landscape architects based in Seattle, and in October 2000 they won the competition – Piet's first commission in the United States. From then, until its completion in 2004, this ambitious urban renewal project, a 1 hectare (2.5 acre) roof garden on top of an underground car park and disused railway, was never far from Piet's thoughts or his drawing board.

In September 2001, Piet was in Chicago with John Coke when the devastating attack on the Twin Towers took place. With all flights cancelled, they decided to visit nurseryman Neil Diboll, whom Piet had met years earlier, at Prairie Nursery in Wisconsin. They also met Roy Diblik at Northwind Perennial Farm, who gave them a tour of the Schulenberg Prairie in Illinois, one of the world's oldest tall-grass prairie restorations [11]. These meetings with kindred spirits were the start of lasting friendships.

The following year, Piet was in the US again and the Battery Conservancy in New York asked Piet to produce a horticultural plan for the 'Gardens of Remembrance' in the Battery, a public park named after the seventeenth century artillery batteries that were part of defences on the southern most tip of Manhattan (see pp.66–7). This was followed by a richly planted 1.6 hectare (4 acres) grove of plane trees, now called the Battery Bosque, which opened in 2005. Anja accompanied Piet to Chicago, in 2002, where they visited the Gensburg-Markham Prairie, a vast stretch of protected land south of the city, bursting with wildlife and plant diversity, which had once been used as a resting place on cattle drives to Chicago. The unique plant communities of these prairies inspired Piet to adapt his planting plans for both the Lurie Garden and The Battery in New York. By using a rich palette of North American perennials, he was 'able to achieve durability in the planting scheme and open visitors' eyes to the beauty of their native plants'.

Also in 2002, the nursery at Hummelo celebrated its twentieth anniversary, and Piet was awarded the Veitch Memorial Medal by the Royal Horticultural Society, an honour reserved for people of any nationality who have made an outstanding contribution to the science and practice of horticulture. I had stepped down as editor of *Gardens Illustrated* and had not long been in the post of Curator at the Chelsea Physic Garden, where Piet and Anja paid regular visits. The plants in the systematic order beds, modelled on the botanic gardens of Carolus Clusius (1526–1609) in Leiden, were of particular interest to Piet, who would cast a discerning eye over any unusual plants.

While Piet was working at the Chelsea Flower Show in 2000, he met the British garden designer Tom Stuart-Smith, who invited him to collaborate on a major restoration of the historic garden at Trentham, near Stoke-on-Trent in Staffordshire, in the north of England. There Piet designed 'The River of Grass', which lived up to its name, and 'The Floral Labyrinth', as well as the long borders that flank Stuart-Smith's redesign of the Italian Garden. When the gardens were opened in 2004 they were described by garden designer James Alexander-Sinclair as 'the garden make-over of the decade'. Simultaneously, Piet was working on the redesign of a 1.6 hectare (4 acre) walled garden at Scampston Hall in North Yorkshire, which opened in 2005. There he created a contemporary garden that included a perennial-rich meadow in what had once been an eighteenth-century kitchen garden [10]. Piet's intention was to encourage biodiversity and reduce the maintenance required in a traditional vegetable garden, although it does, like all his gardens, require maintenance.

Thus the new millennium brought success and recognition for Piet and Anja, but there was sadness too. In 2003 Jelena de Belder died, followed in 2005 by the irrepressible Rob Leopold, leaving Jasper Helmantel and Jojanneke Bijkerk to take over the nursery at Cruydt-Hoeck. At Leopold's funeral, Henk Gerritsen said, 'We are gathered together to witness Rob's transition to another gradient. He will continue to inspire us.' Gerritsen was at that time ill himself, and he died in 2008, just before his book *Essay on Gardening* was published. Piet found a new literary partner in Noel Kingsbury, an established

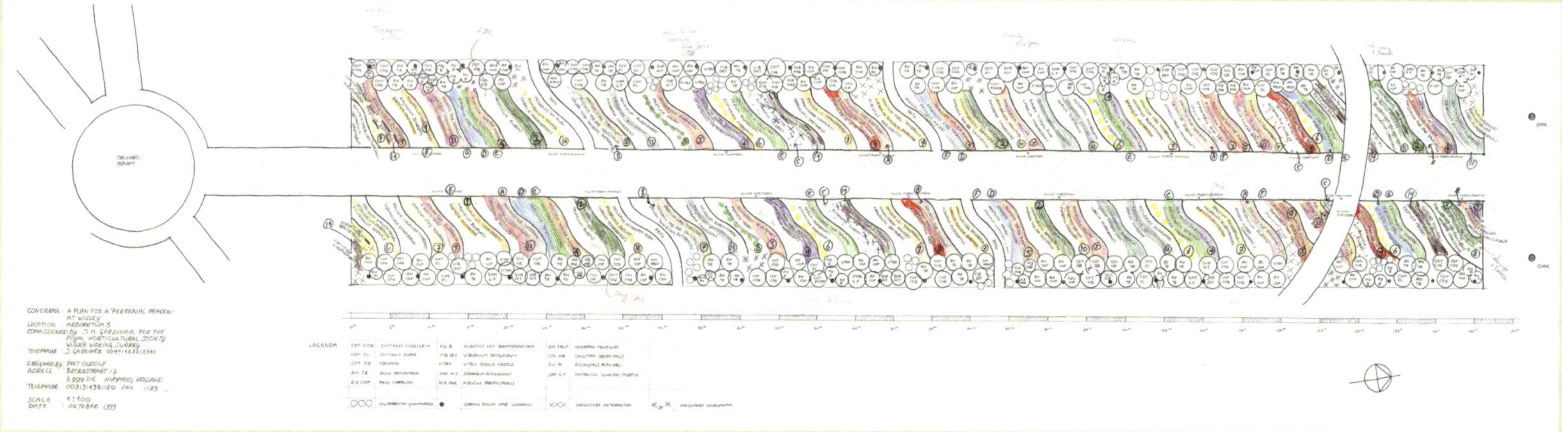

7

8

9

10

11

[7] Plan for perennial borders at RHS Wisley, Surrey, UK, 1999; [8] Dream Park, Enköping, Sweden, 1996; [9] Piet and Anja with Henk Gerritsen; [10] 'Drifts of Grass' garden at Scampston Hall's Walled Garden in North Yorkshire, UK, 2005; [11] Schulenberg Prairie on the west side of The Morton Arboretum, Lisle, Illinois.

12

13

British garden writer and plantsman who first visited Hummelo in 1994. He has since produced five books with Piet.

In 2004 Piet had been approached by the American landscape architecture firm James Corner Field Operations, along with architects Diller Scofidio + Renfro, to collaborate on a project to transform a 2.4 kilometre (1.5 mile) stretch of disused, elevated railway track that had been used to transport freight into New York City (pp.82–95). This was the beginning of the world-famous linear park called the High Line, which opened in 2009 (a second section opened in 2011, a third in 2014 and the 'Spur' in 2019). The High Line was the perfect project for Piet's plant-led design, building on his knowledge of American native flora. Then in 2007, while working on the High Line, Piet was contacted by Gross Max, an award-winning firm of European landscape architects based in Edinburgh, to work with them on a redesign for Potters Field Park, beside Tower Bridge on the River Thames, in London.

With so many of his projects taking place abroad, Piet often combined his trips with speaking engagements, or with his position as visiting professor at Harvard Graduate School of Design. After much deliberation, Piet and Anja decided to close the nursery in 2011, and in that year work began on a new purpose-built studio, designed by architect Hein Tomesen (p.10). The new studio overlooks what was the nursery, which has been replanted with a layered matrix of plants that requires minimum maintenance.

While Piet is famous for transforming public spaces, he also undertakes private commissions and smaller projects. I have a personal fondness for a small, ephemeral garden he created with the Swiss architect Peter Zumthor at the Serpentine Gallery in London in 2011 [12], one of several commissions for museums and art galleries. Their 'Hortus Conclusus' was a jewel, the result of another perfect partnership. At the time, Piet was also working on the planting for the Queen Elizabeth Olympic Park, which opened in London for the 2012 Summer Olympic Games (pp.124–29). That same year, the Royal Institute of British Architects made Piet an Honorary Fellow, and in 2013 he received the Prince Bernhard Culture Fund Award, the premier Dutch prize.

Soon after working on the Serpentine Gallery exhibit, Piet was approached by the Swiss art dealer Iwan Wirth and his wife, Manuela, who asked him to create a garden around their new contemporary art gallery, Hauser & Wirth, at Durslade Farm in Bruton, Somerset. This horticultural work of art, in which 0.6 hectares (1.5 acres) is filled with 25,000 perennials, was acknowledged as a masterpiece when it opened in 2014 (pp.130–49). This was the start of an ongoing collaboration, as the couple later commissioned Piet to create the gardens for Chillida Leku museum in Hernani, Spain (pp.192–95) and for a new arts centre in the repurposed outbuildings of an eighteenth-century naval hospital in Illa del Rei, Menorca, which opened in 2021 (pp.234–43).

Most garden designers work in multi-disciplinary practices, but Piet has always preferred to work alone in his studio, collaborating externally with teams of multi-skilled professionals. Like all gardeners, he believes to share is to multiply. 'The public projects allow me to share my work with a large number of people, while the private gardens are, well, private,' he says. He likes to think that all his gardens are a promise for the future, and 'like a good marriage', he says, 'I want my planting designs to work together as they age.' In 2017 the American filmmaker Thomas Piper made a documentary called *Five Seasons: The Gardens of Piet Oudolf*, which captures how Piet creates seasonal interest with plant compositions on the New York High Line, at the Hauser & Wirth garden in Somerset, and at Hummelo [13].

In 2018 Piet completed the planting in the new sculpture garden at the Singer Museum in Laren, east of Amsterdam (pp.166–71). The museum and concert hall are an extension of a private house called Villa de Wilde Zwanen ('Wild Swans'), built in 1901 by William Singer, son of an American steel baron. The garden, open to the public, is the perfect foil for sculptures by contemporary Dutch artists that have a connection with nature. That year Piet was awarded the Singer Prize by the Singer Memorial Foundation, where he was described as an artist who uses the earth as his canvas.

Before Covid struck in early 2020, Piet finished in 2019 a garden for the Maggie's Centre (which provides support for cancer patients and their families) at the Royal Marsden Hospital in Sutton, outside London (pp.180–85). The Delaware Botanic Gardens in the US also opened their gates to one of Piet's iconic perennial meadow gardens, thanks to 50 volunteers who planted 17,000 plants out of a total of 64,000, in just five days (pp.162–65).

As travel became increasingly difficult during the pandemic, Piet was kept busy in his studio. With the help of German landscape architect Bettina Jaugstetter, he was able to complete the planting at the Vitra Campus, in Weil am Rhein, Germany (pp.202–19). This ever-changing but precisely composed wilderness provides the perfect setting for a unique collection of exhibition spaces and architect-designed buildings. The garden, which fully opened in the summer of 2021, is divided into four planting concepts, linked by sinuous paths, that allow visitors to lose themselves in the rich perennial planting, rather than just pass through it.

Work also continued on the Oudolf Garden Detroit on Belle Isle, in the mouth of the Detroit River [14]. This 1 hectare (2.5 acre) garden (pp.226-33) came about when a member of the Garden Club of Michigan wrote a fan letter, asking Piet to design a garden for Detroit. In this garden, which opened in August 2021, curious visitors can identify individual plants by scanning QR codes.

In 2021 Piet received an Honorary Royal Designer for Industry award from the Royal Society of Arts in London, and he continues to work on projects around the world, large and small [16, 17]. His distinctive planting plans and drawings, informed by years of close observation, remain integral to his working process and are now considered works of art. He also keeps in touch with thousands of Instagram followers, sharing new projects and changes in the garden at Hummelo.

In March 2022 Piet was back in England after nearly four years, visiting RHS Wisley to develop plans to revise the borders that lead to the glasshouse that he created over 20 years ago (see pp.254-55). He also walked the route of what will one day be the Camden Highline in London [15]. This project, running along a disused railway viaduct from Camden Gardens in the west to near King's Cross station in the east, will connect communities and increase access to green space. Piet will again be working with James Corner Field Operations, as he did on the Olympic Park in 2012 and the High Line in New York from 2004 to 2019. Simon Pitkeathley, CEO of Camden Highline, said, 'What started as a rather mad idea is going forward with the finest team I could ever have hoped to be working with.'

Piet has become the leading figure in a movement that promotes a more natural and resilient approach to urban landscape design. His understanding of how plants behave in different soils and climates makes his designs both distinctive and sustainable. As the gardening writer and historian Tim Richardson says, 'Piet is an original thinker and a proselytiser for an attitude to planting rather than a "look".' His work embraces and accepts the natural process of the life and death of plants and recognizes how plants adapt to change, which is something we will all have to do if humankind is to preserve a peaceful and sustainable way of life in the future.

14

15

16

17

[12, previous page] Serpentine Gallery Pavilion, designed with Peter Zumthor, summer 2011; [13, previous page] Film still from *Five Seasons: The Gardens of Piet Oudolf* by Thomas Piper, 2017; [14] Planting of the Oudolf Garden Detroit on Belle Isle, a state park on the Detroit River, 2021; [15] Site of the Camden Highline in London; [16] Piet's studio at Hummelo, designed by Hein Tomesen; [17] Piet in the garden at Hauser & Wirth Menorca.

PLANTING DESIGN FOR THE
NEW BORDERS AT THE GLASSHOUSE
RHS WISLEY WOKING
DATE: JUNE 2021
SCALE 1:100
0 2 4 6 8 10M

lans for the new planting at RHS Wisley, n Surrey, which will incorporate a variety of tyles. The hierarchy of paths (wide and main, arrow and subsidiary) plays an important role in many naturalistic planting projects. The smaller paths invite an intimate exploration, whereas the wider ones encourage taking a wider perspective and view of the garden.

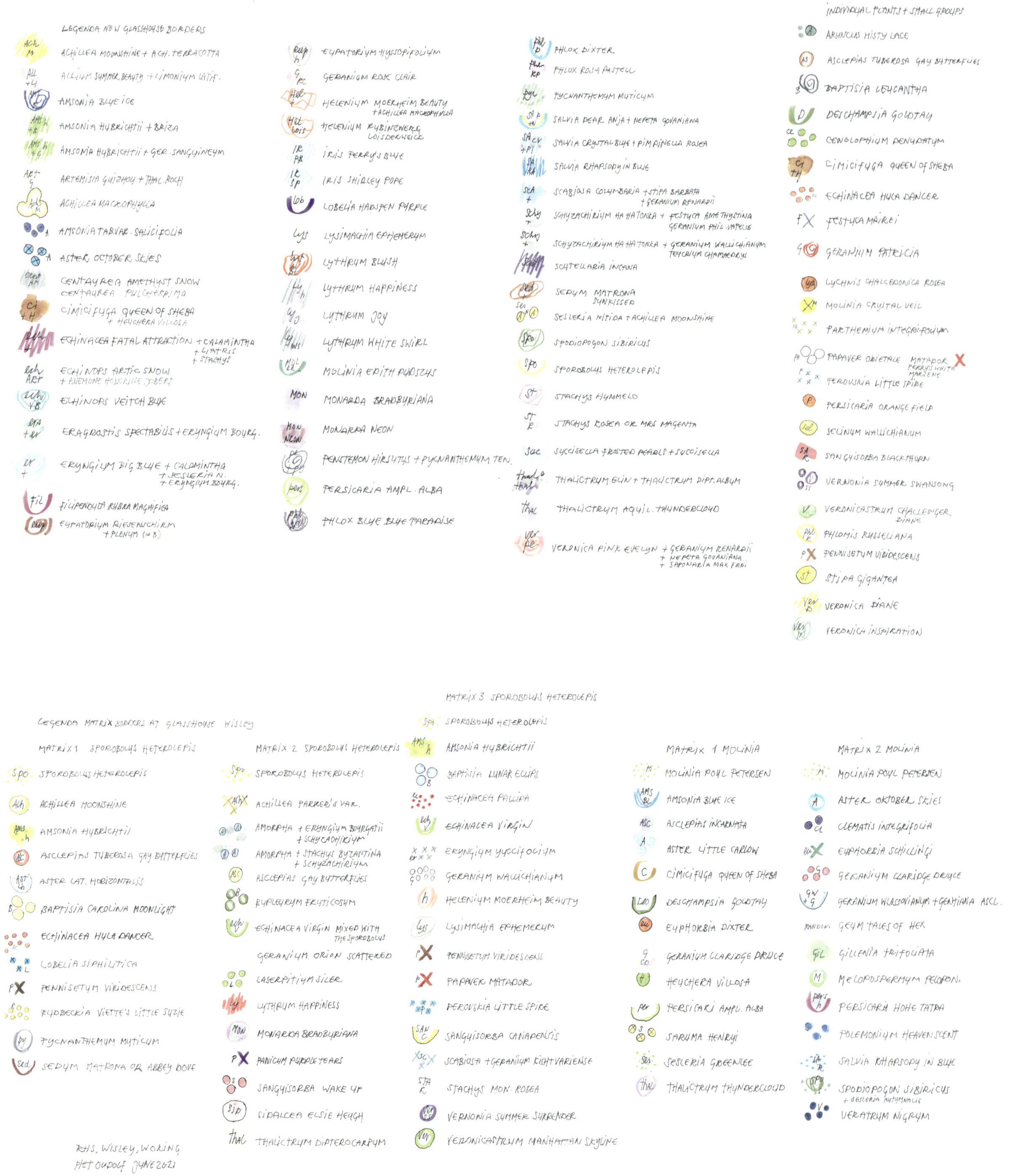
LEGENDA NEW GLASSHOUSE BORDERS
ACHILLEA MOONSHINE + ACH. TERRACOTTA
ALLIUM SUMMER BEAUTY + LIMONIUM LATIF.
AMSONIA BLUE ICE
AMSONIA HUBRICHTII + BRIZA
AMSONIA HUBRICHTII + GER. SANGUINEUM
ARTEMISIA GUIZHOU + THAL. ROCH
ACHILLEA MACROPHYLLA
AMSONIA TAB.VAR. SALICIFOLIA
ASTER OCTOBER SKIES
CENTAUREA AMETHYST SNOW
CENTAUREA PULCHERRIMA
CIMICIFUGA QUEEN OF SHEBA
+ HEUCHERA VILLOSA
ECHINACEA FATAL ATTRACTION + CALAMINTHA
+ LIATRIS
+ STACHYS
ECHINOPS ARTIC SNOW
+ ANEMONE HONORINE JOBERT
ECHINOPS VEITCH BLUE
ERAGROSTIS SPECTABILIS + ERYNGIUM BOURG.
ERYNGIUM BIG BLUE + CALAMINTHA
+ SESLERIA N.
+ ERYNGIUM BOURG.
FILIPENDULA RUBRA MAGNIFICA
EUPATORIUM RIESENSCHIRM
+ PLENUM (IN B)
EUPATORIUM HYSSOPIFOLIUM
GERANIUM ROSE CLAIR
HELENIUM MOERHEIM BEAUTY
+ ACHILLEA MACROPHYLLA
HELENIUM RUBINZWERG
LOISDERWEICK
IRIS PERRY'S BLUE
IRIS SHIRLEY POPE
LOBELIA HADSPEN PURPLE
LYSIMACHIA EPHEMERUM
LYTHRUM BLUSH
LYTHRUM HAPPINESS
LYTHRUM JOY
LYTHRUM WHITE SWIRL
MOLINIA EDITH DUDSZUS
MONARDA BRADBURIANA
MONARDA NEON
PENSTEMON HIRSUTUS + PYCNANTHEMUM TEN.
PERSICARIA AMPL. ALBA
PHLOX BLUE BLUE PARADISE
PHLOX DIXTER
PHLOX ROSA PASTELL
PYCNANTHEMUM MUTICUM
SALVIA PEAR ANJA + NEPETA GOVANIANA
SALVIA CRYSTAL BLUE + PIMPINELLA ROSEA
SALVIA RHAPSODY IN BLUE
SCABIOSA COLUMBARIA + STIPA BARBATA
+ GERANIUM RENARDII
SCHYZACHIRIUM HA HA TONKA + FESTUCA AMETHYSTINA
GERANIUM PHIL. VAPELLE
SCHYZACHIRIUM HA HA TONKA + GERANIUM WALLICHIANUM
TEUCRIUM CHAMAEDRYS
SCUTELLARIA INCANA
SEDUM MATRONA
SUNKISSED
SESLERIA NITIDA + ACHILLEA MOONSHINE
SPODIOPOGON SIBIRICUS
SPOROBOLUS HETEROLEPIS
STACHYS HUMMELO
STACHYS ROSEA OR MRS MAGENTA
SUCCISELLA FROSTED PEARLS + SUCCISELLA
THALICTRUM ELIN + THALICTRUM DIPT. ALBUM
THALICTRUM AQUIL. THUNDERCLOUD
VERONICA PINK EVELYN + GERANIUM RENARDII
+ NEPETA GOVANIANA
+ SAPONARIA MAX FREI
INDIVIDUAL PLANTS + SMALL GROUPS
ARUNCUS MISTY LACE
ASCLEPIAS TUBEROSA GAY BUTTERFLIES
BAPTISIA LEUCANTHA
DESCHAMPSIA GOLDTAU
CENOLOPHIUM DENUDATUM
CIMICIFUGA QUEEN OF SHEBA
ECHINACEA HULA DANCER
FESTUCA MAIREI
GERANIUM PATRICIA
LYCHNIS CHALCEDONICA ROSEA
MOLINIA CRYSTAL VEIL
PARTHENIUM INTEGRIFOLIUM
PAPAVER ORIETALE MATADOR
PERRY'S WHITE
MARLENE
PEROVSKIA LITTLE SPIRE
PERSICARIA ORANGE FIELD
SELINUM WALLICHIANUM
SANGUISORBA BLACKTHORN
VERNONIA SUMMER SWANSONG
VERONICASTRUM CHALLENGER
DIANE
PHLOMIS RUSSELIANA
PENNISETUM VIRIDESCENS
STIPA GIGANTEA
VERONICA DIANE
VERONICA INSPIRATION
LEGENDA MATRIX BORDERS AT GLASSHOUSE WISLEY
MATRIX 1 SPOROBOLUS HETEROLEPIS
SPOROBOLUS HETEROLEPIS
ACHILLEA MOONSHINE
AMSONIA HUBRICHTII
ASCLEPIAS TUBEROSA GAY BUTTERFLIES
ASTER LAT. HORIZONTALIS
BAPTISIA CAROLINA MOONLIGHT
ECHINACEA HULA DANCER
LOBELIA SIPHILITICA
PENNISETUM VIRIDESCENS
RUDBECKIA VIETTE'S LITTLE SUZIE
PYCNANTHEMUM MUTICUM
SEDUM MATRONA OR ABBEY DOVE
MATRIX 2 SPOROBOLUS HETEROLEPIS
SPOROBOLUS HETEROLEPIS
ACHILLEA PARKER'S VAR.
AMORPHA + ERYNGIUM BOURGATII
+ SCHYZACHIRIUM
AMORPHA + STACHYS BYZANTINA
+ SCHYZACHIRIUM
ASCLEPIAS GAY BUTTERFLIES
BUPLEURUM FRUTICOSUM
ECHINACEA VIRGIN MIXED WITH
THE SPOROBOLUS
GERANIUM ORION SCATTERED
LASERPITIUM SILER
LYTHRUM HAPPINESS
MONARDA BRADBURIANA
PANICUM PURPLE TEARS
SANGUISORBA WAKE UP
SIDALCEA ELSIE HEUGH
THALICTRUM DIPTEROCARPUM
MATRIX 3 SPOROBOLUS HETEROLEPIS
SPOROBOLUS HETEROLEPIS
AMSONIA HUBRICHTII
BAPTISIA LUNAR ECLIPS
ECHINACEA PALLIDA
ECHINACEA VIRGIN
ERYNGIUM YUCCIFOLIUM
GERANIUM WALLICHIANUM
HELENIUM MOERHEIM BEAUTY
LYSIMACHIA EPHEMERUM
PENNISETUM VIRIDESCENS
PAPAVER MATADOR
PEROVSKIA LITTLE SPIRE
SANGUISORBA CANADENSIS
SCABIOSA + GERANIUM KISHTVARIENSE
STACHYS MON. ROSEA
VERNONIA SUMMER SURRENDER
VERONICASTRUM MANHATTAN SKYLINE
MATRIX 1 MOLINIA
MOLINIA POUL PETERSEN
AMSONIA BLUE ICE
ASCLEPIAS INCARNATA
ASTER LITTLE CARLOW
CIMICIFUGA QUEEN OF SHEBA
DESCHAMPSIA GOLDTAU
EUPHORBIA DIXTER
GERANIUM CLARIDGE DRUCE
HEUCHERA VILLOSA
PERSICARI AMPL. ALBA
SARUMA HENRYI
SESLERIA GREENLEE
THALICTRUM THUNDERCLOUD
MATRIX 2 MOLINIA
MOLINIA POUL PETERSEN
ASTER OKTOBER SKIES
CLEMATIS INTEGRIFOLIA
EUPHORBIA SCHILLINGI
GERANIUM CLARIDGE DRUCE
GERANIUM WLASSOVIANUM + GENTIANA ASCL.
RANDOM GEUM TALES OF HEX
GILLENIA TRIFOLIATA
MELOPOSPERMUM PELOPON.
PERSICARIA HOHE TATRA
POLEMONIUM HEAVEN SCENT
SALVIA RHAPSODY IN BLUE
SPODIOPOGON SIBIRICUS
+ SESLERIA AUTUMNALIS
VERATRUM NIGRUM
RHS, WISLEY, WOKING
PIET OUDOLF JUNE 2021

THE PLANTS

Plants are characters that I compose with. I put plants on stage and let them perform...

Piet Oudolf

Plant Directory: Perennials, Trees and Shrubs

Here we present the raw materials: plants of the Oudolf signature style. This selection is designed to show the range of plants Piet uses and to illustrate the type and character of the species and varieties that give his planting its impact. His focus is overwhelmingly on perennials of temperate climates, but he also uses trees and shrubs to create frameworks for the perennial plantings, or for the garden or landscape as a whole (see the list of trees and shrubs on p.279). Flowering times are for European conditions at the latitude of the Netherlands. A number of his recent projects have involved creating designs for dry summer climates, which of necessity need a different plant palette; see the list of Mediterranean plants on p.279.

Particularly important is the appearance of the plant after flowering; while many perennials look untidy or simply fail to contribute to a planting composition, many of these will have good seedheads for several months. RHS AGM Plants of Merit are noted – these have been granted an Award of Garden Merit by the UK's Royal Horticultural Society if they have proved exceptional performance in trials. The common names given are those which have some heritage and common usage, and are overwhelmingly those of British or North American native species. Names used for marketing purposes or convenience have not been used. The use of the scientific names is highly recommended to avoid confusion.

POSITION	SIZE/HEIGHT		FLOWERING SEASON (NORTHERN HEMISPHERE)		APPEARANCE OUTSIDE FLOWERING SEASON	HARDINESS NOTE
Sun		cm in			Foliage interest	All these plants are hardy in the Netherlands (USDA Zone 8, with a minimum of -12°C / 10°F) and the overwhelming majority are hardy in a North American Midwest climate (e.g. Chicago, Zone 5, minimum -29°C / -20°F).
Partial shade	Very tall	> 200 cm > 79 in	Jan	Jul	Distinctive foliage before flowering	
Shade	Tall	140–200 cm 55–79 in	Feb	Aug	Little interest outside flowering season	
	Medium	80–140 cm 31–55 in	Mar	Sep	Post-flowering interest	
	Short	30–80 cm 12–31 in	Apr	Oct	Good seedhead interest for 1–2 months after flowering	
	Low	< 30 cm < 12 in	May	Nov	High-quality seedhead interest until well into winter	
			Jun	Dec		

60 cm
24 in
Jun–Jul

ACHILLEA 'MOONSHINE'

Yarrow. Grey-green foliage and a relatively long flowering season. Good for dry situations and more reliable than many of the warm-coloured *Achillea*. Dense, slowly expanding clump.

70 cm
27 in
Jun–Aug

ACHILLEA 'WALTHER FUNCKE'

Yarrow. A vibrantly coloured plant, good for dramatic interest. Expanding clump, but often breaks up, so not always persistent; best on light soils.

90 cm
35 in
Jun–Sept

ACHNATHERUM CALAMAGROSTIS

An attractive grass that forms a dense, non-expanding clump, integrating well with flowering perennials.

180 cm
71 in
Sep–Oct

ACONITUM CARMICHAELII WILSONII GROUP

Monkshood. A good later-season blue. Erect habit, slowly expanding clumps, with solid flower heads. RHS AGM Award of Garden Merit.

180 cm
71 in
Sep–Oct

ACTAEA 'QUEEN OF SHEBA'

Bugbane. A tall, clump-forming, deciduous perennial with dark purple leaves, upright arching stems with scented flowers. Forms tight, non-spreading clump. Long-lived.

40 cm
16 in

ADIANTUM PEDATUM

Five-fingered maidenhair fern. Delicate foliage is a good contrast for broad leaves of flowering perennials. Slowly spreading, long-lived.

40 cm 16 in | Jul–Aug

ALLIUM 'SUMMER BEAUTY'

Usefully later-flowering than other alliums. A tight, slowly expanding clump rather than individual bulbs.

50 cm 20 in | Jun–Jul

AMSONIA 'BLUE ICE'

Bluestar. Clusters of blue, star-shaped flowers in late spring–early summer and good yellow-orange autumn colour. Dense, slowly expanding clumps. Slow to establish, but long-lived.

70 cm 27 in | Jun–Jul

AMSONIA HUBRICHTII

Arkansas bluestar. The best yellow autumn colour of any perennial. Dense, slowly expanding clumps. Slow to establish, but long-lived.

90 cm 35 in | Jun–Jul

AMSONIA TABERNAEMONTANA *VAR.* SALICIFOLIA

Eastern bluestar. Good yellow-orange autumn colour. Dense, slowly expanding clumps. Slow to establish, but long-lived.

80–100 cm 31–39 in | Aug–Sep

ANEMONE HUPEHENSIS 'HADSPEN ABUNDANCE'

Japanese anemone. Bold, divided foliage. Slow to establish, but robust and long-lived, potentially out-competing smaller plants. Strong clump-forming habit.

150 cm 59 in | Aug–Oct

ANEMONE × HYBRIDA 'HONORINE JOBERT'

Japanese anemone. Bold, divided foliage. Slow to establish, but robust and long-lived, potentially out-competing smaller plants. Strong clump-forming habit.

175 cm / 69 in | Jul–Aug

ARTEMISIA LACTIFLORA 'GUIZHOU GROUP'

White mugwort. Divided aromatic dark green foliage on upright, dark stems. Tight clump. Long-lived.

120 cm / 47 in | Jun–Jul

ARUNCUS 'HORATIO'

Goat's beard. Attractive divided foliage with wide-spreading stems. Tight clump, but slowly expanding. Long-lived.

150 cm / 59 in | Jul–Sep

ASCLEPIAS INCARNATA

Swamp milkweed. Upright habit. Tight clump, short-lived. May self-seed. Naturally of moister soils. Good seedheads.

80 cm / 31 in | Jun–Aug

ASCLEPIAS TUBEROSA

Butterfly weed. Famous as an important food source of monarch butterflies. Upright habit. Tight clump, may be short-lived. May self-seed. Naturally of drier soils. Remarkable seedheads.

50 cm / 20 in | Jul–Aug

ASTILBE CHINENSIS 'VISION IN PINK'

Chinese astilbe. Vigorous plant with steadily spreading clumps. Distinctive divided foliage. Long-lived. Does not need as much moisture as other *Astilbe*. Good winter skeleton.

60 cm / 24 in | Jun–Jul

ASTRANTIA MAJOR 'ROMA'

Masterwort. One of many very similar varieties. Slowly expanding clumps. Long-lived and intolerant of heat and drought.

100 cm 39 in | May–Jun

BAPTISIA 'CAROLINA MOONLIGHT'

A striking plant with attractive foliage and an almost shrub-like bulk. Slowly clump-forming, long-lived.

120 cm 47 in | Jun–Jul

BAPTISIA LEUCANTHA

Largeleaf wild indigo. Attractive foliage and an almost shrub-like bulk. Slowly clump-forming, long-lived.

100 cm 39 in | Jul–Oct

BISTORTA AMPLEXICAULIS 'ORANGE FIELD' (FORMERLY PERSICARIA)

Very useful plant for filling space. Steadily expanding clump of weed-suppressing foliage. Very long flowering season, like all varieties of the species.

40 cm 16 in | Apr–May

BRUNNERA MACROPHYLLA

Siberian bugloss. Large, distinctive, rounded leaves make an impact in summer. Steadily expanding clump. Multiples can be used for ground cover.

160 cm 63 in | Jun–Jul

CALAMAGROSTIS × ACUTIFLORA 'KARL FOERSTER'

Feather reed grass. Exceptionally useful for its vertical presence and robust weatherproof qualities over a long season. Slow-spreading. Attractive until late winter.

30 cm 12 in | Aug–Oct

CALAMINTHA NEPETA *SUBSP.* NEPETA

Lesser catmint. Tight clumps produce sprawling foliage that is very useful for gap-filling among upright perennials or grasses. Strongly aromatic.

60 cm 24 in | Jun–Jul

CAMPANULA GLOMERATA 'CAROLINE'

Clustered bellflower. Flowers atop upright stems. Tight clumps, usually long-lived. Good on shallow, calcareous soils.

40 cm 16 in

CAREX MONTANA

Mountain sedge. Tight clumps of neat deciduous foliage. *Carex* is a large genus of low grassy plants; there are many other very similar species.

35 cm 14 in

CAREX MORROWII 'MOSTEN'

Japanese sedge. Tight clumps of neat evergreen foliage. There are many other very similar varieties, some with variegated foliage.

80–120 cm 31–47 in | Jun–Jul

CENOLOPHIUM DENUDATUM

Baltic parsley. A delicate member of the cow parsley family, flowering for longer than most. Needs cool, moist conditions. RHS AGM Award of Garden Merit.

70 cm 28 in | Jun–Aug

DESCHAMPSIA CESPITOSA 'GOLDTAU'

Tufted hair grass. A fine-textured grass, very useful for mixing with perennials. Slowly expanding clumps, generally long-lived. Good on poor soils.

120 cm 47 in | Jul–Aug

ECHINACEA PALLIDA

Pale purple coneflower. With its reflexed petals, an attractive alternative to the familiar *E. purpurea*. More robust, with a longer lifespan. A bee and butterfly magnet.

80 cm 31 in | Jul–Sep

ECHINACEA PURPUREA 'FATAL ATTRACTION'

Purple coneflower. Large purple-pink blooms with an orange centre; very popular and with a great many similar varieties. All have limited lifespans, although self-seeding is sometimes possible.

70 cm 28 in | Jul–Sep

ECHINACEA PURPUREA 'VIRGIN'

Purple coneflower. The best white. Tight clumps, non-spreading. All *Echinacea* dislike being crowded by other plants.

120 cm 47 in | Jun–Jul

ECHINOPS RITRO 'VEITCH'S BLUE'

Small globe thistle. Dramatic, dark, thistle-like foliage with perfectly spherical flower and seedheads. Forms tight, non-spreading clump. Intensely coloured flower heads. All *Echinops* have a lifespan of more than five years, often more than ten.

40 cm 16 in | Mar–Apr

EPIMEDIUM GRANDIFLORUM

Violet-flowered barrenwort. A highly effective deciduous foliage plant for shade, contrasting well with other woodland plants. Long-lived, non-spreading clumps.

50 cm 20 in | Jun–Jul

ERAGROSTIS SPECTABILIS

Love grass. Cloud-like flower- and seedheads contrast effectively with more clearly defined plant forms. Tight, non-spreading clump. Can self-seed.

50 cm 20 in | Jun–Jul

ERYNGIUM BOURGATII

Smaller than most *Eryngium*. Tight clump of attractive foliage, tolerant of summer drought and a range of soil types.

100 cm 39 in | Jul–Aug

ERYNGIUM YUCCIFOLIUM

Rattlesnake-master. Dramatic sculptural plant with a long season of interest. Non-spreading, but can seed strongly. Many other *Eryngium* similar.

80 cm 31 in | Jul–Aug

ERYNGIUM × ZABELII 'BIG BLUE'

Distinct form combined with an intense colour. Long-lived tight clumps. Good cut flower.

120 cm 47 in | Sep–Oct

EUPATORIUM HYSSOPIFOLIUM

Hyssopleaf thoroughwort. White flowers useful for planting in drifts among other species. Fine foliage. A good pollinator plant for drier soils. Self-seeds.

80 cm 31 in | May–Jun

EUPHORBIA GRIFFITHII 'DIXTER'

Spurge. Unusual flower colour early in the year. Spreading to form extensive clumps, but not strongly competitive with other plants.

100 cm 39 in | Aug–Sep

EURYBIA × HERVEYI 'ASTER TWILIGHT'

Hervey's aster. Vibrant flowers on strongly spreading plant that fills in gaps quickly.

160–250 cm 63–98 in | Jul–Sep

EUTROCHIUM MACULATUM 'RIESEN-SCHIRM' (FORMERLY EUPATORIUM)

Joe Pye weed. Statuesque plant, popular with pollinators. Forms a long-lived, tight clump. Windproof when established, best in moister soil.

80 cm / 31 in · Jun-Jul

FESTUCA MAIREI

Atlas fescue. Tough, drought-tolerant grass that forms attractive non-spreading tussocks. Good for dotting among other dry-habitat perennials.

180 cm / 71 in · July

FILIPENDULA RUBRA 'VENUSTA'

Queen of the prairie. Impressive plant for damp soil, steadily spreading into extensive, loose clumps. Good for wild garden situations. Attractive in autumn, with chestnut-brown flower skeletons.

50 cm / 20 in · Jun-Sep

GERANIUM × OXONIANUM 'CLARIDGE DRUCE'

Cranesbill. Repeat-flowering plant that forms steadily spreading clumps. Can be untidy post-flowering. Very useful for weed suppression and rough places, and for naturalizing in grass. Can self-seed.

60-90 cm / 24-35 in · May-Jun

GERANIUM PHAEUM 'RAVEN'

Dusky cranesbill. Mysteriously toned flowers above steadily spreading clumps. Can self-seed. Very good for naturalizing.

25-50 cm / 10-20 in · Jun-Jul

GERANIUM SANGUINEUM 'ALBUM'

Bloody cranesbill. One of many colour forms of this neat, low-growing and very usefully drought-tolerant cranesbill. Good on calcareous soils and as ground cover between taller perennials.

50 cm / 20 in · Jun-July

GERANIUM 'SIRAK'

Cranesbill. Steadily spreading clumps of foliage, which colours well in the autumn. RHS AGM Award of Garden Merit. One of many hybrid varieties of *Geranium*.

50 cm 20 in | Apr–Jun

GEUM 'TALES OF HEX'

Avens. Slowly spreading clumps of neat, low foliage, useful for front-of-border situations. There is a growing number of new varieties of this genus, all with 'hot' flower colours. All have potential as ground cover.

100 cm 39 in | Jun–Jul

GILLENIA TRIFOLIATA

Bowman's root. Upright stems forming a bushy plant, which can be spectacular in flower. Good orange autumn colour. Non-spreading, but long-lived.

40 cm 16 in | Aug–Sep

HAKONECHLOA MACRA

Japanese forest grass. Neat foliage, with a 'combed' appearance, makes this a very attractive plant. Steadily spreading to form extensive mats. Intolerant of drought.

140 cm 55 in | Jul–Aug

HELENIUM 'LOYSDER WIECK'

Sneezeweed. A vigorous, upright perennial that forms clumps with time. It needs fertile, relatively moist soil.

120 cm 47 in | Jul–Aug

HELENIUM 'MOERHEIM BEAUTY'

Sneezeweed. A vigorous, upright plant, clump-forming. Needs fertile, moist soil and protection from slugs. An old variety, raised by the father of Mien Ruys. There are many varieties, all in 'warm' colours.

40 cm 16 in | Aug–Sep

HEUCHERA VILLOSA

Alum root. Grown primarily for its luxuriant foliage, this impressive plant is clump-forming and far more reliable than other *Heuchera* varieties.

45 cm / 18 in · Jul–Sep

HYLOTELEPHIUM TELEPHIUM 'SUNKISSED' (FORMERLY SEDUM)

Orpine. One of many varieties of a very useful group of plants. Much appreciated by pollinators, especially butterflies. Dense clump, long-lived.

100 cm / 39 in · June

IRIS SIBIRICA

Siberian iris. Short flowering season, but has other distinctive qualities. Tolerant of a wide range of conditions. Forms very long-lived, expanding clumps.

100 cm / 39 in · Sep–Oct

KIRENGESHOMA PALMATA

Yellow wax-bells. A distinguished-looking plant that forms steadily expanding, long-lived clumps. Does well in hot, humid summers, but dislikes drought.

100 cm / 39 in · Jul–Aug

LIATRIS SPICATA

Button snakewort. Fine linear foliage. Forms slowly expanding clumps. Tolerant of drought, and best on drier soils.

60 cm / 24 in · Jul–Aug

LIMONIUM PLATYPHYLLUM (SYN. LIMONIUM LATIFOLIUM)

Broad-leaved statice. Large, glossy leaves in loose rosettes from a long-lived, very slowly expanding clump. Very heat- and drought-tolerant.

80 cm / 31 in · Jul–Aug

LOBELIA SIPHILITICA

Blue cardinal flower. A short-lived but often self-seeding plant that adds a fine splash of blue to moist environments. Upright habit.

☀ ◐ | 80 cm 31 in | Apr–June | ✳

LUNARIA REDIVIVA

Perennial honesty. A long season of scented flowers is followed by unusual seedheads. Forms a small clump, long-lived – very distinct from the familiar annual honesty.

☀ | 140 cm 55 in | Jun–Jul

LYCHNIS CHALCEDONICA 'ROSEA'

Flower of Bristol. Striking flowers atop very upright stems. Forms a small clump and often self-seeds. Can be grown in grass, as it does well with a hay cut.

☀ | 100 cm 39 in | Jul–Aug

LYTHRUM VIRGATUM 'JOY'

Purple loosestrife. Delicate flowers on upright stems. Non-spreading, but reasonably long-lived.

☀ | 180 cm 71 in | Aug–Oct | ✳

MISCANTHUS SINENSIS 'FLAMINGO'

Chinese silver grass. Bold clumps of foliage that steadily spread, more rapidly in warmer climates. Good winter appearance. One of many varieties.

☀ | 200 cm 79 in | Aug–Oct | ✳

MISCANTHUS SINENSIS 'MALEPARTUS'

Chinese silver grass. One of the taller varieties. All are seen at their best when backlit in low winter sunlight. Increasingly regarded as invasive in some warmer parts of North America.

☀ ◐ | 220 cm 87 in | Jul–Oct

MOLINIA CAERULEA SUBSP. ARUNDINACEA 'TRANSPARENT'

Purple moor-grass. Very airy stems with cloud-like flower/seedheads, which last longer on the plant than most other cultivars. Can self-seed.

100 cm 39 in | Jul-Oct

MOLINIA CAERULEA *SUBSP.* CAERULEA 'MOORHEXE'

Purple moor-grass. Upright habit and good autumn colour, stems breaking off before mid-winter. Good for mass effects. Like all *Molinia*, it forms a non-expanding, long-lived clump.

220 cm 87 in | Jul-Oct

MOLINIA LITORALIS 'DARK BEAUTY'

Purple moor-grass. Tall grass. Orange-purple stems in autumn with very dark seedheads.

60 cm 24 in | May-Jun

MONARDA BRADBURIANA

Eastern beebalm. Whorled flowers attract many bees. Clump-forming, and tending to be more reliable and less disease-prone than other *Monarda* varieties.

120 cm 47 in | Jul

MONARDA 'NEON'

Beebalm. Strongly aromatic, the brightest of many varieties in pink, red and purple shades. Quickly forms a loose, spreading clump.

120 cm 47 in | Aug-Oct

PANICUM VIRGATUM 'SHENANDOAH'

Switch grass. Clump-forming grass with fine flower/seedheads, and one of the best of many varieties for autumn colour.

50 cm 20 in | May-Jun

PAPAVER ORIENTALE 'MATADOR'

Oriental poppy. Dramatic flowers produced early, then dormant in summer. An ideal 'scatter plant'. Long-lived, slowly expanding clump.

80 cm 31 in | Jun–Aug

PENSTEMON DIGITALIS 'HUSKER RED'

Beardtongue. Almost evergreen foliage, with a strong red tinge to younger growth. Forms a long-lived, tight clump. Hardy, unlike many *Penstemon.*

80 cm 31 in | Jun–Jul

PHLOMIS RUSSELIANA

Turkish sage. Exceptionally valuable plant with evergreen foliage and very robust seedheads. Can be used as ground cover, as it spreads steadily.

120–40 cm 47–55 in | May–Jun

PHLOMIS TUBEROSA

Distinctive flower heads and foliage, useful for being one of the few tall perennials for early summer.

160 cm 63 in | Jul–Aug

PHLOX PANICULATA 'DIXTER'

A favourite out of the vast number of varieties of this popular scented summer flower. Strong-growing, forming clumps.

100 cm 39 in | May–June

POLYGONATUM × HYBRIDUM 'WEIHENSTEPHAN'

Solomon's seal. A distinguished-looking plant for shade. Can be dormant in summer. Eventually spreads to form large but loose clumps.

60 cm 24 in

POLYSTICHUM SETIFERUM

Soft shield fern. This evergreen fern and its various cultivars are some of the most tolerant of drier soils and stronger light conditions. Long-lived, dense clump.

90 cm 35 in | Aug-Sep

PYCNANTHEMUM MUTICUM

Short-toothed mountain mint. Intensely aromatic foliage on a strongly spreading perennial, the main interest of which is the silvery bracts. One of the best for pollinators.

80 cm 31 in | Jun-Jul

RODGERSIA PINNATA 'SUPERBA'

Grown primarily for foliage, this is one of many distinguished-looking plants for moist places. Slow to establish, but eventually spreading.

120 cm 47 in | Jul-Sep

RUDBECKIA SUBTOMENTOSA 'LITTLE HENRY'

Sweet coneflower. Upright habit with striking, small flower petals. Very slowly spreading, long-lived. There are several other *Rudbeckia*, all with late golden flowers.

120 cm 47 in | Aug-Sep

SALVIA 'LITTLE SPIRE' (FORMERLY PEROVSKIA)

Russian sage. Tight clump of semi-woody upright growth, best cut back in late winter after enjoying its ghostly white stems.

40 cm 16 in | May-Jun

SALVIA HYBR. 'RHAPSODY IN BLUE'

Forms slowly expanding, long-lived clumps. Other cultivars offer shades of purple, some pink, some white. All are notably drought-tolerant.

160 cm 63 in | Jun-Aug

SANGUISORBA 'BLACKTHORN'

Burnet. Very elegant divided foliage before flowering. Damp soils preferred. Long-lived tight clump.

180 cm
71 in
Aug-Sep

SANGUISORBA CANADENSIS

White burnet. Stands better than many varieties. Upright growth best seen above surrounding plants. Long-lived tight clump.

100 cm
39 in
May-Jun

SANGUISORBA MENZIESII 'WAKE UP'

Burnet. Attractive foliage, as valuable early in the season as the vivid flowers are later.

180 cm
71 in
Jul-Aug

SANGUISORBA TENUIFOLIA

Slender-leaved burnet. Elegantly narrow divided foliage. Damp soils preferred. Long-lived tight clump, inclined to flop.

80 cm
31 in
Sep-Oct

SCHIZACHYRIUM SCOPARIUM 'HA HA TONKA'

Little bluestem. A distinctive cultivar of a key North American native grass. This variety has hairy grey leaves and does not flop.

90 cm
35 in
Aug-Sep

SCUTELLARIA INCANA

Hoary skullcap. Upright habit. Slowly spreading, long-lived clump. Like many other flowers with bilateral symmetry, much appreciated by wild bees.

120 cm
47 in
Jul-Oct

SELINUM WALLICHIANUM

Wallich milk parsley. Much later than most of the cow parsley family, with very elegant filigree foliage. Can self-seed.

40 cm 16 in | Aug-Sep

SESLERIA AUTUMNALIS

Autumn moor-grass. Semi-evergreen foliage forming steadily expanding mats. Ideal for matrix planting, as it is not too competitive.

40 cm 16 in | Aug-Sep

SESLERIA 'GREENLEE HYBRID'

Moor-grass. More glaucous foliage on a stronger-growing plant than *Sesleria autumnalis*. All *Sesleria* are good on poorer, drier soils. Tight growth helps suppress weeds.

80 cm 31 in | Jul-Sep

SPOROBOLUS HETEROLEPIS

Prairie dropseed. An amazingly adaptable, long-lived grass, the fine flower/seedheads of which have an attractive scent. Dense, non-competitive clumps. Can be slow to establish. Spectacular autumn colour.

70 cm 28 in | Jun-Jul

STACHYS MONIERI 'ROSEA'

Like several related *Stachys*, the foliage is attractive before flowering. Forms dense low clumps. Long-lived.

200 cm 79 in | Jun-Aug

STIPA GIGANTEA

An evergreen grass forming tight, non-expanding clumps. Dramatic when backlit against a dark background. Drought-tolerant and very long-lived.

80-100 cm 31-39 in | Jul-Aug

SUCCISA PRATENSIS

Devil's bit scabious. Tight clumps of shiny leaves produce clouds of flowers that are very popular with pollinators. Needs group planting to make an impact.

120 cm 47 in | Sep–Oct

SYMPHYOTRICHUM 'LITTLE CARLOW' (FORMERLY ASTER)

Aster. A truly superb aster, with sprays of pale blue flowers above a slowly expanding clump. Low-maintenance and reliable. Long-lived.

160 cm 63 in | Sep–Oct

SYMPHYOTRICHUM NOVAE-ANGLIAE (FORMERLY ASTER)

New England aster. Upright habit. Slowly expanding clump. There are many different cultivars in shades from dark purple to pale blue, pink and white.

90 cm 35 in | Oct–Nov

SYMPHYOTRICHUM OBLONGIFOLIUM 'OCTOBER SKIES' (FORMERLY ASTER)

Aster. An aster with a bushy habit, in contrast to the somewhat 'leggy' appearance of many. Slowly expanding, long-lived clump.

120–150 cm 47–59 in | May–Jun

THALICTRUM AQUILEGIFOLIUM

Meadow rue. One of the few tall perennials for late spring. Flower colours vary from pink-purple to white. Tight clump, reasonably long-lived.

140–200 cm 55–79 in | Jul–Sep

THALICTRUM DELAVAYI 'ALBUM'

Chinese meadow rue. The purity of the white of the flowers makes this a useful cultivar. Slowly expanding clump. Has light growth that tends to be inconspicuous after flowering.

180 cm 71 in | Jul–Aug

THALICTRUM ROCHEBRUNEANUM

Lavender mist meadow rue. All *Thalictrum* have attractive, finely divided foliage that is a feature pre-flowering. All grow best in cool summer climates.

80 cm / 31 in · Sep-Oct

TRICYRTIS FORMOSANA

Toad lily. *Tricyrtis* are rare examples of late-flowering shade plants. This is relatively vigorous, forming a long-lived, spreading clump with time.

90 cm / 35 in · Sep-Oct

TRICYRTIS 'SHIMONE'

Toad lily. *Tricyrtis* have glossy leaves on upright stems. Most flourish in humid summer climates and may not be tolerant of drier soils.

120 cm / 47 in · Jul-Aug

VERATRUM NIGRUM

Black false hellebore. An exceptionally distinguished plant, valuable for a long season of structure. Can be slow to establish, but long-lived. Beware of slugs!

120 cm / 47 in · Sep-Oct

VERNONIA 'SUMMER SWANSONG'

Ironweed. Ironweeds are useful for the very end of the gardening year, much loved by pollinators. This cultivar is more compact than most. Tight, long-lived clump.

100 cm / 39 in · Jun-Jul

VERONICA LONGIFOLIA 'EVELINE'

Garden speedwell. A longer-lived and more compact variety of a perennial noted for its colourful upright flower spikes. Slowly expanding, dense, long-lived clumps.

140 cm / 55 in · Jul-Aug

VERONICASTRUM VIRGINICUM 'CHALLENGER'

Culver's root. An exceptionally good plant for long-season interest, now with many cultivars. Dense, very slowly expanding, long-lived clump.

SHRUBS

Amelanchier lamarckii Snowy mespilus
Amorpha fruticosa Bastard indigo
Callicarpa dichotoma Purple beautyberry
Calycanthus 'Hartlage Wine' Sweetshrub
Chaenomeles 'Moerloosii' Japanese quince
Chaenomeles 'Toyo Nishiki' Japanese quince
Clerodendron trichotomum Harlequin glorybower
Corylopsis pauciflora Buttercup witch hazel
Cotinus coggygria Smoke tree
Crataegus prunifolia Broad-leaved cockspur thorn
Cydonia oblonga Quince
Fothergilla major Mountain witch alder
Hamamelis 'Diane' Witch hazel
Hamamelis 'Pallida' Witch hazel
Hamamelis vernalis Witch hazel
Hydrangea aspera Rough-leaved hydrangea
Hydrangea macrophylla
Hydrangea quercifolia Oak-leaved hydrangea
Ilex verticillata Winterberry
Indigofera kirilowii
Lindera salicifolia Spicebush
Lonicera fragrantissima Winter-flowering honeysuckle
Paeonia lutea Tree peony
Philadelphus coronarius Variegated mock orange
Rhododendron species Rhododendron
Rhus copallinum Dwarf sumac
Salix purpurea 'Nancy Saunders' Willow
Sambucus 'Guincho Purple' Elder
Syringa 'James McFarlane' Lilac
Syringa patula 'Miss Kim' Lilac
Vaccinium corymbosum Blueberry
Viburnum × bodnantense 'Dawn' Arrowwood
Viburnum nudum Withe rod
Viburnum plicatum 'Mariesii' Japanese snowball
Viburnum prunifolium American black haw
Vitex negundo var. *cannabifolia* Chaste tree

TREES

Acer campestre Field maple
Acer japonicum Japanese maple
Acer triflorum Three-flowered maple
Aesculus parviflora Bottlebrush buckeye
Betula nigra 'Heritage' Black birch
Carpinus betulus Hornbeam
Cercidiphyllum japonicum Katsura
Cercis 'Forest Pansy' Redbud
Chionanthus retusus Chinese fringe tree
Cornus 'Stellar Pink' Dogwood
Cornus kousa var. *chinensis* Chinese dogwood
Cornus mas Cornelian cherry
Euonymus planipes Flat-spiked spindle
Halesia monticola Caroline silverbell
Maackia amurensis
Magnolia 'Leonard Messel' Magnolia
Magnolia 'Yellow Butterfly' Magnolia
Magnolia virginiana Magnolia
Malus hupehensis Hupeh crab
Nyssa sylvatica Tupelo
Parrotia persica Persian ironwood
Pinus mugo Dwarf mountain pine
Pinus sylvestris Scots pine
Prunus 'Okamé' Cherry
Prunus yedoensis Yoshino cherry
Sassafras albidum Sassafras
Stewartia monadelpha
Styrax japonicus Japanese snowbell
Styrax obassia Fragrant snowbell

PLANTS SUITABLE FOR MEDITERRANEAN OR OTHER DRY SUMMER CLIMATES

Achillea species; see p.261
Agapanthus species African lily
Ballota acetabulosa Greek hoarhound
Convolvulus cneorum Silverbush
Dianthus anatolicus
Dietes grandiflora Fairy iris
Echium species Viper's bugloss
Echium virescens
Eragrostis spectabilis Love grass; see p.266
Eryngium species; see pp.266–67
Euphorbia characias Spurge
Euphorbia dendroides
Euphorbia rigens
Gaura lindheimeri White gaura
Geranium 'Rozanne'
Geranium sanguineum Striped bloody cranesbill; see p.268
Lavandula angustifolia English lavender
Lavandula stoechas French lavender
Leonotis leonurus Lion's tail
Limonium perezii
Liriope species
Melianthus major Honey flower
Muhlenbergia capillaris Pink muhly grass
Origanum 'Herrenhausen' Oregano
Perovskia species (now Salvia); see p.274
Phlomis fruticosa Jerusalem sage
Phlomis russeliana Turkish sage; see p.273
Rosmarinus officinalis Rosemary
Salvia fruticosa
Salvia lycioides Autumn sage
Santolina Lavender cotton
Saponaria × olivana Cushion soapwort
Sedum species White stonecrop
Stachys byzantina Lamb's ear
Teucrium chamaedrys Wall germander
Thymus vulgaris Thyme
Tulbaghia violacea Society garlic

List of Gardens

1982–
Hummelo
Hummelo, Gelderland, Netherlands
With Anja Oudolf

1992 and 2002
Floriade, International Horticultural Exhibition
Netherlands
Masterplan: Niek Roozen
Collaborators: Jacqueline van der Kloet, Hein Koningen, Climmy Schneider

1995–96
Drömparken (Dream Park)
Enköping, Sweden
Commissioned by Stefan Mattson

1996–97
Bury Court
Hampshire, UK
Commissioned by John Coke

2001–04
Lurie Garden in Millennium Park
Chicago, Illinois, USA
Landscape Architect: Gustafson Guthrie Nichol, Seattle with Lighting Designer Robert Israel
Project Architect: Terry Guen
Collaborators: Roy Diblik, Jacqueline van der Kloet
Head Horticulturist: Jennifer Davit, Kathryn Deery, Laura Ekasetya and Colleen Lockovitch

2003
The Battery, Gardens of Remembrance and the Bosque (2003–05) and Bikeway Gardens (2010–12)
New York City, New York, USA
Landscape Architect: Saratoga Associates and Quennell Rothschild & Partners
Collaborators: WXY architecture + urban design

2004
Scampston Hall and Walled Garden
Malton, North Yorkshire, UK
Commissioned by Legard Family

2006–
The High Line
New York City, New York, USA
Commissioned by Friends of the High Line and the City of New York
Landscape Architect: James Corner Field Operations (Project Lead)
Architect: Diller Scofidio + Renfro
Collaborators: Joshua David (Co-founder), Robert Hammond (Co-founder), Friends of the High Line
Directors of Horticulture: Melissa Fisher, Richard Hayden, Andi Pettis, Eric Rodriguez, Tom Smarr

2008/2011
Skärholmens Perennpark
Sweden
Commissioned by Stefan Mattson
Head Gardener: Anneli Wallgren

2008–
Private Garden
Nantucket Island, Massachusetts, USA
Masterplan: James Corner Field Operations
Head Gardener: Steve Thompson (d.2021)

2008–12
Berne Park: Sedimentation Tanks (Theatre of Plants)
Bottrop, Germany
Berne Park Masterplan: Davids Terfrüchte + Partner
Landscape Architect: GROSS MAX with Piet Oudolf

2010
Leuvehoofd
Rotterdam, Netherlands
Landscape Architect: DS+V

2010
Ichtushof
Rotterdam, Netherlands
Landscape Architect: Erick van Egeraat
Architect: Erick van Egeraat

2010
Westerkade
Rotterdam, Netherlands
Landscape Architect: Okra

2010
'Il Giardino delle Vergini' at the Venice Biennale
Venice, Italy
Commissioned by the Venice Biennale
Architect: Kazuyo Sejima of SANAA
Collaborator: Climmy Schneider

2011
Hortus Conclusus, Serpentine Gallery Pavilion (Temporary Installation), London, UK
Commissioned by Serpentine Gallery
Architect: Peter Zumthor

2011
Private Roof Garden
New York City, New York, USA
Landscape Architect: Richard Herbert

2011–18
(original plantings, 2011; long border added in 2018)
Maximilian Park
Hamm, Germany
Landscape Architect: Petra Rieke

2012–14
Queen Elizabeth Olympic Park
London, UK
Masterplan: LDA Design with Hargreaves Associates
Landscape Architect: James Corner Field Operations

2012
Palais de Tokyo
Paris, France
Private Commission

2012–14
Oudolf Field at Hauser & Wirth Somerset
Durslade Farm, Somerset, UK
Project Landscape Architect: Urquhart & Hunt
Architect: Luis Laplace
Head Gardener: Mark Dumbelton

2013
Roof Garden at David Zwirner Gallery
New York City, New York, USA
Architect: Selldorf Architects

2013–14
De Vlinderhof
Máximapark, Utrecht, Netherlands
Initiated by Marc Kikkert
Masterplan Máximapark: West 8
Garden Stewardship: Team of volunteers

2014–16
Private Garden
Near Omaha, Iowa, USA
Collaborators: Roy Diblik, Austin Eischeid, Adam Woodruff

2015
Garden for Japan Garden & Flower Show (Temporary Installation)
'Huis ten Bosch' Theme Park,
Nagasaki, Japan
Landscape designer: Yuko Nagamura

2016–
Private Garden
Berkshire, UK
Project Landscape Architect: Urquhart & Hunt
Collaborators: Chris and Toby Marchant

2017–19
Private Garden
Amsterdam, Netherlands
Project Architect: Deltavormgroep

2017–19
Meadow Garden at Delaware Botanic Garden
Pepper Creek, Dagsboro, Delaware, USA
Landscape Architect: Rodney Robinson of Robinson Anderson Summers, Inc.
Head Horticulturist: Stephen Pryce Lea
Garden Stewardship: Team of volunteers

2018
Singer Laren
Laren, Netherlands
Project Architect: Deltavormgroep
Collaborator: Climmy Schneider

2018
Private Garden
Pennsylvania, USA

2019
Noma
Copenhagen, Denmark
Landscape Architect: Thing Brandt
Head Gardeners: Liv Linea Holm, Kirsty Marchant, Naja Jessen
Collaborators: Climmy Schneider, Tom de Witte and team of volunteers

2019
Maggie's at the Royal Marsden Hospital
London, UK
Architect: Ab Rogers Design
Collaborators: Chris and Toby Marchant; Steve Swatton Landscapes

2019
Museo Chillida Leku
Hernani, Spain
Architect: Luis Laplace
Project Landscape Architect: Álvaro de la Rosa Maura
Collaborator: Tom de Witte

2020
Private Garden
Basel, Switzerland
Project Architect: Deltavormgroep, PG landscape architects
Collaborator: Bettina Jaugstetter

2020
Oudolf Garten on the Vitra Campus
Weil am Rhein, Germany
Project Architect: Deltavormgroep
Collaborators: Bettina Jaugstetter, Cassian Schmidt

2020–21
Private Garden
Nieuw Haamstede, Zeeland, Netherlands
Collaborators: Linda van de Lavoir, Tom de Witte
Head Gardener: Stuart Huizen

2021
Oudolf Garden, Belle Isle
Detroit, Michigan, USA
Initiated by the Garden Club of Michigan with Maura Campbell and Jean Hudson
Landscape Architect: InSite Design Studio
Project Management: Meredith Simpson, Duncan Campbell
Collaborators: Austin Eischeid, Roy Diblik, Laura Ekasetya
Head Horticulturist: Richard Thomas
Garden Stewardship: Oudolf Garden Detroit (all-volunteer group)

2021
Oudolf Garden at Hauser & Wirth Menorca
Illa del Rei, Menorca, Spain
Architect: Luis Laplace
Project Architect: Álvaro de la Rosa Maura
Collaborator: Tom de Witte

In addition, the following are acknowledged for their role in commissioning or collaborating on various projects:

Beatrice and Charles Brown
Dudok Family
Rolf Fehlbaum
Kurt Fortin
Judith Giesberts
Sigrid Gray
Barbara Katz
Gert-Jan van der Kolk
Rasa Laurinaviciene
Mireia Massagué
Hans Ulrich Obrist
Anja Oudolf
Hanna Packer
Julia Peyton-Jones
Warrie Price
René Redzepi
Mar Rescalvo Pons
Raymond Sander and Sheryl Swed
Bob and Polina Schlott
Tony Spencer
Gregg Tepper
Kate Le Vasseur
Iwan and Manuela Wirth
Alice Workman
Peter Zumthor

Bibliography

Darke, Rick, Piet Oudolf and Lorraine Ferguson. *Gardens of the High Line: Elevating the Nature of Modern Landscapes*. Portland, OR: Timber Press, 2017.

Dusoir, Rory. *Planting the Oudolf Gardens at Hauser & Wirth Somerset*. With a foreword by Piet Oudolf. London: Filbert Press, 2019.

Gerritsen, Henk, and Piet Oudolf. *Dream Plants for the Natural Garden*. London: Frances Lincoln, 2000 and revised edition 2013; originally published in Dutch as *Méér Droomplanten*: Terra, 1999.

King, Michael, and Piet Oudolf. *Gardening with Grasses*. London: Frances Lincoln, 1998; originally published in Dutch as *Prachtig gras*: Terra, 1996.

Kloe, Fabian de, Peter Veenstra and Joep Vossebeld, eds. *Landscape Works with Piet Oudolf and LOLA: In Search of Sharawadgi*. Rotterdam: nai010, 2021.

Oudolf, Piet, and Henk Gerritsen. *Planting the Natural Garden*. Portland, OR: Timber Press, 1st edition, 2003, 2nd edition, 2019; simultaneously published in Dutch in 2019 as *Droomplanten*: HL Books. Originally published in Dutch as *Droomplanten*: Terra, 1990 and 2003.

Oudolf, Piet, and Noel Kingsbury. *Designing with Plants*. London: Conran Octopus, 1999.

Oudolf, Piet, and Noel Kingsbury. *Landscapes in Landscapes*. New York: Monacelli Press, 2011; simultaneously published in Dutch as *Landschap in landschap*: Terra.

Oudolf, Piet, and Noel Kingsbury. *Oudolf | Hummelo: A Journey Through a Plantsman's Life*. New York: Monacelli Press, 1st edition, 2015, 2nd edition, 2021; simultaneously published in Dutch as *Oudolf | Hummelo*: HL Books.

Oudolf, Piet, and Noel Kingsbury. *Planting: A New Perspective*. Portland, OR: Timber Press, 2013; simultaneously published in Dutch as *Plannen en planten – Een nieuw perspectief*: Terra Lannoo.

Oudolf, Piet, and Noel Kingsbury. *Planting Design: Gardens in Time and Space*. Portland, OR: Timber Press, 2005; simultaneously published as *Planten voor morgen – Ontwerpen met visie*.

Index

Page numbers in *italics* refer to illustrations

Acknowledgements

First and foremost, my thanks go to Anja – my constant companion throughout the years.

I would also like to thank the following people for their contribution to the making of this book:

Mark Ashbee
Rosie Atkins
Edwin Barendrecht
Jonny Bruce
Victoria Clarke
James Corner
Rick Darke
Laura Ekasetya
Jennifer Harkins
Julia Hasting
Walter Herfst
Liv Linea Holm
Jason Ingram
Bettina Jaugstetter
Naja Jessen
David Jiménez
Noel Kingsbury
Hélène Lesger
Kirsty Marchant
Hans Ulrich Obrist
Cassian Schmidt
Tino Sehgal
Max Shackleton
Ariane Spanier
Claire Takacs
Tuula Whitlow

–*Piet Oudolf*

Publisher's Acknowledgements

With thanks to the following for their help in the making of this book: Caitlin Arnell Argyles, Adela Cory, Rosie Fairhead, Diane Fortenberry, Joanne Glättli, João Mota, Tracey Smith, Hans Stofregen, Matilda Southern-Wilkins and David Tibbs. We are also very grateful to the garden owners, head gardeners, museums and institutions for all their support and help with permissions and pictures.

Picture Credits

All images courtesy and © Piet Oudolf, unless listed below:

Photograph by Mark Ashbee: 11, 13, 16, 253.16; © Edwin Barendrecht: 150-51, 154, 155; Photograph by Taylor Beaufore: 233t; © Richard Bloom: 142-3; Courtesy Camden Highline: 253.15; Photograph by Rick Darke/© Simone Leigh, Courtesy Matthew Marks Gallery: 90b; GAP Photos: Richard Bloom 263bm, Tim Gainey 268bl, Nova Photo Graphik 266bm, 274tl, Jan Smith 270br, Nicola Stocken 265br; Courtesy Hauser & Wirth: 'Elogio del vacío VI' (2000) by Eduardo Chillida on Hauser & Wirth Menorca/© Zabalaga-Leku, DACS, London 2022. Courtesy the Estate of Eduardo Chillida and Hauser & Wirth/Photo by Daniel Schäfer 234-5, 'Le Père Ubu' (1974) by Joan Miró on Hauser & Wirth Menorca/© Successió Miró/ADAGP, Paris and DACS London 2022/Private Collection. Courtesy Hauser & Wirth/Photo by Daniel Schäfer 236-7, Photo by Daniel Schäfer 242tl, 242tm, 242tr, 242br, 243; © Walter Herfst: 98-9, 100bl, 100-1bm, 101br, 102-3, 104, 107, 166-7, 170, 250.12; Photograph by Liv Linea Holm: 172-3, 178b, 179; © Jason Ingram: 130-31, 132-3, 145, 146, 147, 156-7, 249.10, 261bl, 264bm, 271br, 272tl, 273tm, 275br, 276tm; Photograph by Bettina Jaugstetter/Cassian Schmidt: 216br, 217, 265bl; Photograph by Naja Jessen: 178tr; Photo © David Jiménez: 192-3, 195; © Ricardo Labougle: 253.17; Courtesy Lurie Garden: 81; Courtesy Maggie's at the Royal Marsden/Photograph by Jennifer Harkins: 180-81, 184t, 185; Photograph by Kirsty Marchant: 178tl; Film still from *Five Seasons: The Gardens of Piet Oudolf*, directed by Thomas Piper: 250.13; Photograph by Timothy Schenck/Courtesy Friends of the High Line: 88-9; Photograph by Sevenfotos/Toni Glauser: 196-7, 199, 200; Shutterstock/Mark Baldwin: 249.11; Photograph by Meredith Simpson: 233br; © Ryan Southen: 226-7, 232, 253.14; © Claire Takacs: 4-5, 8, 10, 82-3, 90t, 94-5, 148-9, 220-21, 224, 225tl, 225tr, 225b, 261tl, 261bm, 274tr, 274bm, 278bl; © Vitra: 214-15; © Vitra: Photograph Marek Iwicki 202-3, Photograph Dejan Jovanovic 204, 210-11, 216bl, 218, 219, Photograph Julien Lanoo 216t; Photograph by Tuula Whitlow: 160, 161.

'Planting the Vitra Garden' insert: © Vitra/Photograph Dejan Jovanovic: Figures 2, 11; Photograph by Bettina Jaugstetter/Cassian Schmidt: Figures 1, 3, 4, 5, 6, 7, 8, 9, 10, 12.

Phaidon Press Limited
2 Cooperage Yard
London E15 2QR

Phaidon Press Inc.
111 Broadway
NewYork, NY 10006

phaidon.com

First published 2023
Reprinted 2023, 2024, 2025

ISBN 978 1 83866 424 4
ISBN 978 1 83866 682 8 (signed edition)

A CIP catalogue record for this book is available from the British Library and the Library of Congress.

Commissioning Editor: Victoria Clarke
Project Editor: Victoria Clarke with support from Hélène Lesger and Noel Kingsbury
Project Texts and Captions written by Noel Kingsbury
Translation of Introduction by Cassian Schmidt: Joanne Glättli
Production Controller: Adela Cory

Design: Ariane Spanier

Printed in China

Biographies

Piet Oudolf is an influential Dutch garden and landscape designer at the forefront of the New Perennial movement and the author of numerous books on gardening and landscape design. He has designed many residential and commercial gardens, public and private, and his projects can be found throughout The Netherlands, the UK, Ireland, Germany, Sweden and the USA – including the celebrated High Line in New York City, Lurie Garden at Millennium Park in Chicago, Hauser & Wirth gardens in Somerset and Menorca, the sculpture garden at Singer Museum in The Netherlands, and garden installations for the Venice Biennale and the Serpentine Gallery pavilion. He is the recipient of the Gold Medal and title of Best in Show at the Chelsea Flower Show (2000), the Gold Veitch Memorial Medal from the Royal Horticultural Society (2002) and the Award of Distinction from the Association of Professional Landscape Designers in 2010. He was awarded the highest cultural honour in The Netherlands, the Prince Bernhard Culture prize, in 2013.

Contributors

Rosie Atkins FLS was the founding editor of *Gardens Illustrated* magazine and former Curator of the Chelsea Physic Garden. She is Vice-President of the Royal Horticultural Society, Chairman Emeritus of the RHS Woody Plant Committee and in 2024 she received an RHS Elizabeth Medal of Honour (EMH), in recognition of her contribution to horticulture. She is Vice-President of the Metropolitan Public Gardens Association and has served as a trustee of several gardening charities. She is also Chair of Project Giving Back, a charity that aims to fund more than 40 gardens at the Chelsea Flower Show over the coming years in support of good causes.

Jonny Bruce graduated from Cambridge University with a degree in Art History, after which he spent a number of years working at historic properties, including Great Dixter in East Sussex. He subsequently moved to The Netherlands to work at the specialist organic nursery De Hessenhof. There he met Anja and Piet and visited Hummelo on numerous occasions. He currently works as a gardener, writer and design consultant, based in the UK. He is the primary gardener at Prospect Cottage in Dungeness, former home of Derek Jarman.

James Corner is a world-renowned landscape architect and founder of James Corner Field Operations based in New York City, the firm behind the design of New York's High Line, among other important public projects around the world. He is Emeritus Professor of Landscape Architecture and Urbanism at the Weitzman School of Design, University of Pennsylvania, a recipient of the National Design Award, and an elected member of the American Academy of Arts and Letters.

Noel Kingsbury is a planting design consultant and prolific writer on gardens and naturalistic planting, with over 20 books to his name, including the seminal *The New Perennial Garden* (1996) and, most recently, *Wild: The Naturalistic Garden* (2022). He has collaborated with Piet Oudolf on four books, including *Oudolf | Hummelo: A Journey Through a Plantsman's Life* (2015 and 2021). He co-runs the garden education company Garden Masterclass and lectures widely.

Hans Ulrich Obrist is Artistic Director of the Serpentine Galleries in London, Senior Advisor at LUMA Arles, and Senior Artistic Advisor at The Shed in New York. Prior to these, he was Curator of the Musée d'Art de la Ville de Paris. Since his first show, *World Soup (The Kitchen Show)* in 1991, he has curated more than 350 exhibitions.

Cassian Schmidt was formerly Director of Hermannshof Garden, a display and trial garden based on naturalistic perennial plantings at Weinheim, Germany. His research there was focused on the 'New German Style' of planting, using natural plant communities as models for sustainable plant combinations for the urban environment. He is a professor of planting design in the department of landscape architecture at Geisenheim University, and chair of the Working Committee for Perennials of the German Perennial Plant Association.